]

I N choosing global climate change as the subject of our 2007 conference, we knew we would be dealing with a topic fraught with controversy and strong viewpoints. So we made every effort to bring in speakers with a range of viewpoints and expertise. We succeeded in rounding up an impressive array of world-class scholars, and we believe that our contribution to the debate on global warming is unique.

It was one of the rare conferences where both economists and scientists addressed the subject. Much of the public discussion of global warming focuses on strategies to prevent it, but we also brought in speakers who talked about the potential for "mitigation and adaptation." They addressed the science and the economics behind proposals that focus on adjusting to global climate change, rather than preventing it. Other speakers addressed the policy and the politics of global warming. For example, they compared the merits of a carbon tax with a "cap and trade" approach, and took a closer look at the workings of the U.N. and its Intergovernmental Panel on Climate Change (IPCC). The IPCC shared the 2007 Nobel Peace Prize with Al Gore and has become the most prominent source for government policymakers of scientific analysis of climate change. However, one of our speakers argued persuasively that the IPCC's process of review and inquiry is seriously flawed.

We also included a panel of speakers to address a topic that has received little attention from the media: the theology and ethics of global warming. Views on the environment often have theological or spiritual undertones. For some, the guiding principle is God's command in Genesis that man shall "have dominion over the fish of the sea, the birds of the air, and the cattle, and over all the wild animals and all the creatures that crawl on the ground." But many also view Nature as an expression of God's handiwork and still others find in Nature a substitute for formal religion or a belief in God. Both of the latter groups tend to view Nature as something to be protected from mankind. This view, as much as scientific understanding, seems to drive fears and warnings that global warming could be a calamity for the planet and for mankind.

The ethical debate is somewhat different. It highlights the trade-offs involved in proposals to address global warming. For example, is it better to spend more to prevent global warming if that means spending less on

i

reducing world poverty? Would our preferences be any different if global warming were considered to be a natural phenomenon rather than man-made one? The answers are not as simple as the public debate often suggests.

This volume represents a selection of papers presented at the conference. It is not all-inclusive. Some speakers relied on PowerPoint slides that were well-suited to an audience setting but not reproducible in print.

Nonetheless, we believe the papers included here capture the full range of views presented at the conference. For those who wish to see or hear more, our website, www.aier.org, includes selected video and audio clips from all the speakers.

The speakers' participation in our conference did not necessarily imply that they endorse AIER's views or that we endorse theirs. Rather, they accepted our invitation in the spirit of pursuing free inquiry aimed at improving our understanding of global warming.

A final note: The debate over global warming has been tainted on both sides by the influence of money and special interests, or at least the appearance of such influence. In this regard, we are proud to note that AIER does not receive money from governments or corporations, nor do we rely on funding from a few wealthy donors. We do not accept gifts of money or other property that carry with them or include any restrictions on our research. We do not accept advertising, nor do we derive revenue, as many non-profits do, from selling or renting out the names of our supporters. Instead, we seek to derive our support primarily from contributions by the general public and from the distribution of publications representing the results of our scientific research.

All these provisions are set forth in our Bylaws, and all are designed to allow us to conduct genuinely independent research. This was the foremost goal of our founder, Col. E.C. Harwood.

Kerry A. Lynch
Director of Research and Education

Contents

GLOBAL CLIMATE CHANGE

Conference Schedule

Friday, November 2, 2007

Introductory Remarks

Charles E. Murray, President and CEO, AIER
Kerry A. Lynch, Director of Research and Education, AIER
Walker F. Todd, Research Fellow and Conference Organizer, AIER

Session One: 9:00 a.m. to 12:00 noon

Panel One: The Science of Global Climate Change

Moderator: Fred Harwood, Trustee and Secretary of the Corporation, AIER

Carl Wunsch, Massachusetts Institute of Technology
  What is the Current State of Scientific Knowledge about Global Climate Change?
David S. Chapman, University of Utah
  Global Warming: Where Have We Been, and Where are We Going?
William R. Cotton, Colorado State University, *Discussant*

Panel Two: The Scientific Methodology and International Politics of Global Climate Change

Moderator: Walker F. Todd, AIER

David Henderson, University of Westminster
  Governments and Climate Change Issues: A Flawed Consensus
Claudia Rosett, Foundation for Defense of Democracies
  Global Warming and the United Nations
Ross McKitrick, University of Guelph, Ontario, *Discussant*

Lunch: 12:00 noon to 2:00 p.m.
Helen F. Harwood Ballroom, Main Stone House

Session Two: 2:00 to 5:00 p.m.

Panel Three: Scientific Analysis of Climate Change

Moderator: Michael J. Rizzo, Senior Economist, AIER

Gordon E. Michaels, Oak Ridge National Laboratory
*Reducing Energy-Related Greenhouse Gas Emissions:
Nuclear Energy, the Hydrogen Economy, and Global
Fossil-Carbon Management*
William M. Gray, Colorado State University
Hurricane Changes: An Empirical Approach
James I. Mills, University of Utah, *Discussant*

**Panel Four: Rational Economic and Public Policy Responses
to Climate Change in a Globalized World**

Moderator: Walker F. Todd, AIER

Robert O. Mendelsohn, Yale University
Climate Policy: Minimizing Abatement Costs and Climate Damages
Gilbert E. Metcalf, Tufts University
The Carbon Tax Option: A New Look at an Old Idea
Richard L. Stroup, Property and Environment Research Center and
North Carolina State University, *Discussant*

Reception and Dinner: 5:30 to 8:30 p.m.
Helen Harwood Ballroom, Main Stone House

Saturday, November 3, 2007

Session Three: 9:00 a.m. to 12:00 noon

Panel Five: Scientific and Public Policy Choices for Dealing with Climate Change

Moderator: Walker F. Todd, AIER

Richard S. Lindzen, Massachusetts Institute of Technology
  Separating the Core from the Periphery (or Warming from Alarm)
Peter J. Wilcoxen, Maxwell School, Syracuse University
  Economic Policy for Climate Change: Building a Solid Foundation for Critical Investments in New Capital
Kenneth P. Green, American Enterprise Institute, *Discussant*

Panel Six: Ethical and Theological Aspects of Climate Change

Moderator: Michael J. Rizzo, AIER

David Henderson, University of Westminster
  Brief Remarks on Global Salvationism and the Climate Change Debate
E. Calvin Beisner, Knox Theological Seminary
  Some Theological Perspectives on the Climate Change Debate
Robert H. Nelson, School of Public Policy, University of Maryland
  Global Warming and Religion: Climate Policy as Applied Theology
Edward J. Kane, Boston College, *Discussant*

Lunch: 12:00 noon to 2:00 p.m.
Helen F. Harwood Ballroom, Main Stone House

Conference Adjourns

LIST OF PARTICIPANTS

E. Calvin Beisner, Associate Professor of Historical Theology and Social Ethics, Knox Theological Seminary, Fort Lauderdale, Florida. He has been a contributing editor for *World Magazine*, *The Freeman*, *Christian Research Journal*, and *Crosswinds*; author of *Man, Economy, and the Environment in Biblical Perspective*; and coauthor with Ross McKitrick of articles on global climate change. He is also a Fellow of the Acton Institute.

David S. Chapman, Professor of Geology and Geophysics in the College of Mines and Earth Sciences and Dean of the Graduate School at the University of Utah, Salt Lake City. He has taught in Africa, Europe, New Zealand, and North America, and is recognized within his field as one of the top solid-Earth geophysicists for his work on measuring and interpreting geologic heat and mass transfer. His research has covered a wide range of subjects, all concerned with temperatures in the earth, the way in which temperatures influence geologic processes, and measuring the heat loss of the earth.

William R. Cotton, Professor of Atmospheric Science, Colorado State University, Fort Collins. His main field of research has been observation and computer simulation of cumulus clouds and thunderstorms as well as other intermediate-scale cloud systems. Together with Dr. Roger Pielke, Sr., he developed the Regional Atmospheric Modeling System (RAMS). Former editor and co-editor of the *Journal of Atmospheric Sciences*, a fellow of the American Meteorological Society, and author or co-author of four books and over 120 journal articles.

William M. Gray, Professor Emeritus of Atmospheric Science, Colorado State University, Fort Collins. He is the foremost long-range forecaster of hurricane activity in the Atlantic basin. His research involves tropical cyclone genesis, structure changes, intensity variations, and motion. He has received numerous honors and awards from professional societies involved in meteorology and hurricane forecasting and analysis, and has been a fellow of the American Meteorological Society.

Kenneth P. Green, Resident Scholar at the American Enterprise Institute (AEI), Washington, D.C. The subjects of his research include public policy in air pollution and climate change, energy and environment, transport and environment, environmental chemicals, and analysis of

Canadian environmental policy. He has authored numerous policy studies, newspaper and magazine articles, several encyclopedia and book chapters, and a textbook for middle-school students, *Global Warming: Understanding the Debates*. Before joining AEI, Green worked as chief scientist and director of environmental programs at California's Reason Foundation and the Fraser Institute in Vancouver, British Columbia. He was an expert reviewer for the IPCC's 2001 report in Working Group 1.

David Henderson, Visiting Professor in the Business School, University of Westminster, London, U.K. He is a former chief economist of the Organization for Economic Cooperation and Development (OECD). Previously, he worked both as an academic and as a national and international civil servant. Since leaving the OECD, he has been an independent author and consultant, holding visiting appointments in several countries. He works on a range of public policy issues, with particular interest in the balance between liberal or market-oriented tendencies, and regulation and controls in economic policies across the world. His book, *Misguided Virtue*, examines current notions of "corporate social responsibility." In recent years, he has written frequently on the public policy response to the challenges posed by the global warming hypothesis.

Edward J. Kane, James F. Cleary Professor of Finance at Boston College. From 1972 to 1992, he held the Everett D. Reese Chair of Banking and Monetary Economics at Ohio State University. He teaches a course on Ethics in Financial Institution Management at Boston College, and has addressed this topic as a guest speaker at AIER's Summer Fellowship Program. Professor Kane has been a consultant to numerous organizations, including the World Bank, the IMF, the Federal Home Loan Bank Board, and the Federal Reserve System. He is a senior fellow in the Federal Deposit Insurance Corporation's Center for Financial Research, a past president and fellow of the American Finance Association, a former Guggenheim Fellow, and a longtime Research Associate of the National Bureau of Economic Research.

Richard S. Lindzen, Alfred P. Sloan Professor of Meteorology, Department of Earth, Atmospheric, and Planetary Sciences, Massachusetts Institute of Technology. One of the world's leading experts in oceanic cloud formation, he is interested in the broad topics of climate, planetary waves, monsoon meteorology, planetary atmospheres, and hydrodynamic instability. His research involves studies of the role of the tropics in mid-latitude weather and global heat transport, the moisture budget and its role in global change, the origins of ice ages, seasonal effects in atmospheric transport, stratospheric waves, and the observational determination of climate sensitivity. He has developed models for the Earth's climate with specific concern for the stability of the ice caps, the sensitivity to increases in carbon dioxide, the origin of the 100,000-year cycle in glaciation, and the maintenance of regional variations in climate. Professor Lindzen has received honors from the American Meteorological Society and the American Geophysical Union. He is a consultant to the Global Modeling and Simulation Group at NASA's Goddard Space Flight Center and a Distinguished Visiting Scientist at California Institute of Technology's Jet Propulsion Laboratory.

Ross McKitrick, Associate Professor of Economics at the University of Guelph, Ontario, where he specializes in environmental economics. He also is Director of Graduate Studies in the Economics Department at Guelph. His papers on the "hockey stick" model of the paleoclimate, coauthored with Stephen McIntyre, are among those most frequently cited in the literature on global climate change. In 2003 his book, *Taken By Storm: The Troubled Science, Policy and Politics of Global Warming*, coauthored with Christopher Essex, won the Donner Prize for the best book on Canadian Public Policy. He is recognized globally as one of the leading analytical critics of the public policy recommendations arising from mathematical modeling exercises in global climate change. He has published dozens of journal articles on a wide range of topics, including the economic theory of pollution policy, economic growth and air pollution trends, climate policy options, the measurement of temperature and climate change, and statistical methods in paleoclimatology. In 2002, he was appointed as a Senior Fellow of the Fraser Institute in Vancouver, British Columbia.

Robert O. Mendelsohn, Edwin Weyerhaeuser Davis Professor, Professor of Economics, and Professor in the School of Management, Yale University. His Ph.D. dissertation included an integrated assessment model of air pollution that could measure the damages of emissions. In recent years, he has extended this work to greenhouse gases, trying to measure the impact of climate change. Recently, he has returned to studying air pollution in the hope of measuring the marginal damages of emissions across the country. He also has worked on promoting the values of natural ecosystems, including non-timber forest products and ecotourism in tropical rainforests, coral reefs in the Caribbean and Australia, and recreation in the Pacific Northwest and Alaska. Professor Mendelsohn participated in both the Copenhagen Consensus report and the IPCC's Fourth Assessment Report.

Gilbert E. Metcalf, Professor of Economics at Tufts University, and Research Associate at the National Bureau of Economic Research. He also taught at Princeton University and the Kennedy School of Government at Harvard University, and was a Visiting Scholar at the Joint Program on the Science and Policy of Global Change at MIT. He has been a consultant to various organizations, including the Chinese Ministry of Finance, the U.S. Department of the Treasury, and Argonne National Laboratory. His primary research area is applied public finance, with particular interests in taxation, energy, and environmental economics. His current research focuses on policy evaluation and design affecting energy usage and climate change. He has published papers in numerous academic journals, edited two books, and contributed chapters to several books on tax policy.

Gordon E. Michaels, Chief Technology Officer for Energy and Engineering Sciences at the U.S. Department of Energy's Oak Ridge National Laboratory, Tennessee. He was formerly the head of nuclear technology research at Oak Ridge, and has worked at Oak Ridge since 1977.

James I. Mills, Lockheed Martin Science Fellow Emeritus and Adjunct Professor of Architecture and Planning at the University of Utah. He is the president of Sustainable Learning Systems, which provides sustainability consulting and computer simulation services to clients around the world. He spent the first five years of his professional life as a nuclear physicist before switching to global environmental systems research in 1980. In 1997, Dr. Mills organized with Amory Lovins, Paul Hawken, Donella Meadows, and the Rocky Mountain Institute a U.S. Department of Energy-sponsored interdisciplinary forum seeking to characterize and prioritize issues of global sustainability and associated educational imperatives. His current teaching and research interests include urban and natural system simulation, and environmental policy simulation, and creative approaches to the management of "wicked" problems.

Robert H. Nelson, Professor of Environmental Policy at the School of Public Policy of the University of Maryland, College Park. He is a nationally recognized authority on land and natural resource management in the United States, with a particular emphasis on management of federally owned lands. He has written widely on the relationship of culture, religion, and economic policy, including the essay, "Does 'Existence Value' Exist? Environmental Economics Encroaches on Religion," in *Re-Thinking Green: Alternatives to Environmental Bureaucracy* (Robert Higgs and Carl P. Close, editors). His many articles have appeared in the *Journal of Economic Literature*, *Journal of Political Economy*, *Journal of Policy Analysis and Management*, and elsewhere. He is the author of seven books dealing with property rights, zoning, and land-use issues, and was a columnist for *Forbes* in the 1990s.

Claudia Rosett, Journalist-in-Residence with the Foundation for Defense of Democracies, Washington, D.C., writing on international affairs. Over the past 25 years, she has reported from Asia, the former Soviet Union, Latin America, and the Middle East, with a focus on democratic movements and despotic regimes. Since 2002, she has written extensively about the United Nations, including ground-breaking reporting on corruption under the Iraq Oil-for-Food program. Prior to that she spent many years with *The Wall Street Journal* as a staff writer, a member of the Journal's editorial board, a reporter, and then as bureau chief of the Journal's Moscow Bureau, editorial-page editor of *The Asian Wall Street Journal,* and as the Journal's books editor. In 2005 she won both the Mightier Pen Award and the Eric Breindel Award for Excellence in Opinion Journalism for her coverage of the United Nations.

Richard L. Stroup, Senior Associate of the Property and Environment Research Center (PERC), Bozeman, Montana and Professor Emeritus of Economics at Montana State University, Bozeman, where he retired as chairman of the Department of Agricultural Economics and Economics. Currently he is a Visiting Professor at North Carolina State University, Raleigh, and president of the Political Economy Research Institute. Co-author, with James D. Gwartney, of *Economics: Private and Public Choice*, a widely used economics textbook. His book, *Eco-Nomics: What Everyone Should Know about Economics and the Environment*, won the 2004 Sir Anthony Fisher Memorial Award. He is an adjunct scholar at the Cato Institute and a member of the Mont Pelerin Society.

In the Reagan administration, he was director of the Office of Policy Analysis at the U.S. Department of the Interior.

Peter J. Wilcoxen, Associate Professor of Economics and Public Administration, and Director of the Center for Environmental Policy and Administration at the Maxwell School of Syracuse University. He teaches on environmental and resource economics, public economics, taxation, energy policy, managerial economics, and public policy responses to the science of climate change, and has written numerous articles on the economics of climate change. He and Warwick J. McKibbin have offered an alternative model, the McKibbin-Wilcoxen Blueprint, for achieving the reductions of greenhouse gas emissions that were the principal aim of the 1997 Kyoto Protocol. Other recent research initiatives include management of environmental uncertainty and risk by organizations in the Great Lakes region, green building (environmentally friendly residential construction), and improvements in indoor air quality and building energy efficiency.

Carl Wunsch, the Cecil and Ida Green Professor of Physical Oceanography, Department of Earth, Atmospheric, and Planetary Sciences, Massachusetts Institute of Technology. One of the world's foremost experts on the circulation of ocean currents and their effects on ocean climates, he was the leading organizer of ECCO (begun in 1998), an international consortium effort for Estimating the Climate and Circulation of the Ocean. The goal of ECCO is to combine a general circulation model (GCM) with diverse observations in order to produce a quantitative depiction of the time-evolving global ocean state. ECCO is a major field of research for the World Ocean Circulation Experiment (WOCE), of which Professor Wunsch has served as both chairman and co-chairman. He has held numerous visiting professorships, received numerous academic and professional honors, and is the author or co-author of over 225 scientific papers and four books. Professor Wunsch is a member of the American Geophysical Union, the American Meteorological Society, the Oceanography Society, and the Society for Industrial and Applied Mathematics.

INTRODUCTION

Michael J. Rizzo
Senior Economist, AIER

E ARTH has warmed by roughly three-quarters of a degree Celsius since 1880. The coincidence of this warming with a 36 percent increase in atmospheric carbon dioxide concentrations and massive global industrialization has raised fears that human activity is causing global, potentially catastrophic, warming. In an effort to increase our understanding of this complex issue, we invited a group of renowned experts in climate science, economics, and public policy to our November 2007 conference on Global Climate Change. This article presents a summary of the conference; the rest of the book is a collection of selected papers adapted from the speakers' presentations.

Few current issues ignite public passions as much as the subject of global warming. The most widely held view is that 1) it is happening, and 2) we should do something about it, sooner rather than later. Beyond this, however, there is no consensus. The proposed solutions are complex, involve costly trade-offs and sacrifices, and, to be effective, require a degree of genuine global commitment that, experience suggests, would be very hard to achieve. Just as important, the scientific consensus is not as strong as the popular one. The state of scientific knowledge is less certain than the media and others often suggest. On the question of what is going to happen to the climate, many scientists caution that the issue is too complex to give a firm or simple answer.

What Do We Know About the Science of Climate Change?

Climate science is extremely complex for three reasons, according to the climate scientists at our conference — Carl Wunsch (Professor of Oceanography at MIT), David Chapman (Professor of Geology and Geophysics at the University of Utah), William Gray (Professor Emeritus of Atmospheric Science at Colorado State) and Richard Lindzen (Professor of Meteorology at MIT).

First, it is an immature science. The observational records cover a time period that is dwarfed by the relevant geologic time scales. Given the current state of understanding, accurate prediction is not possible.

Second, climate science is virtually the "science of everything." As Professor Lindzen demonstrated, climate damage is not a simple consequence of carbon dioxide (CO_2) emissions and warming. The catastrophic scenarios (ice sheets melting, sea levels rising, starving polar bears, etc.) involve a multitude of factors. Scientists may have an excellent understanding of individual parts of the problem, but no one fully understands how all the parts are linked. To illustrate, predictions of CO_2 emissions are based on models of population growth, economic growth, energy efficiency, energy consumption, technical change, land use, and more. Each of these ties into the climate models and is itself subject to a variety of influences.

Third, modern climate science relies on large, complicated models that seek to describe and predict the climate system, but these models are inherently unreliable and unstable. For instance, climate models predicted three to six times more warming than that which has occurred to date. Complexity does not ensure accuracy. (In this regard, climate models are reminiscent of the failed macroeconomic econometric models that were popular a generation ago.)

Thus, even a broad climate "consensus" may prove fragile. No alternative models were proposed at the conference, and the scientists urged that we take the model results seriously. But they also cautioned that we should prepare for the possibility that the models might miss their targets widely—in *either* direction.

Professor Wunsch expressed concern that society's desire for predictions gives rise to dramatization and dogmatism that hurt the prospects for understanding the underlying science. Professor Gray showed that one can use the data on hurricanes to derive any conclusion desired. He compared hurricane activity from two 50-year periods, 1900-1949 and 1957-2006. During the earlier period, 39 major storms hit land, while during the latter period only 22 major storms landed. Some might conclude that CO_2 leads to fewer storms, because emissions were much higher during the second period. On the other hand, hurricane activity during the past 12 years has far outpaced that of the prior 12 years.

Despite this complexity, there is a vast area where the scientists acknowledged widespread understanding and agreement:

1) Earth has warmed by 0.5 to 0.75 degrees Celsius since 1880. This warming has not been constant across the globe or continuous over time. During some years the temperature has decreased.

2) At 382 parts per million (ppm), atmospheric CO_2 concentrations are

at the highest level of the 150,000-year historical record. That it has taken less than a century to increase from 280 ppm is evidence that human activity has contributed to the increase.

3) CO_2 is a greenhouse gas and therefore could be expected to contribute to some warming, but it is not clear how much.

4) Sea levels have risen by approximately eight inches in a century, and the oceans have become more acidic.

Are Humans Causing the Planet to Warm?

According to the scientists at our conference, identifying mankind's role in the climate question is not easy and might not even be possible for decades. Potential natural sources of climate change include solar variability, aerosol variation, changes in cloud cover, greenhouse gases, changes in land use, and natural variability from factors such as El Niño and the oceans' thermohaline circulation. Changes in any or all of these may be sufficient to account for the changes in global mean temperature since 1880.

Evidence abounds to suggest that CO_2 is far from the most important contributor to global warming. Previous interglacial periods were much warmer and had far lower concentrations of CO_2 in the atmosphere; and it might eventually be demonstrated that CO_2 concentrations *lag* temperature increases. When the temperature change is within a few tenths of a degree, it is difficult to identify the particular causes.

Scientific Options for Dealing with Change

Gordon Michaels (Chief Technology Officer at the Department of Energy's Oak Ridge National Laboratory) demonstrated that the task of reducing CO_2 is formidable. Wind, biomass, solar and other "green" solutions cannot be "scaled up" to meet global energy demands. The best available option for carbon-free energy, he said, is nuclear energy. Using current technology, nuclear energy is only 20 to 30 percent more costly than coal, it is carbon-free aside from the uranium enrichment process, and the supply of uranium-238 that could be used in fast-breeder reactors exceeds *one billion* years. The major obstacle is public resistance. As he put it, every potential storage site for radioactive waste is in some Congressman's district.

Getting from Here to There

Economic problems often boil down to a simple diagnosis: the "price is wrong." In the case of climate change, the zero price of CO_2 is too low. Because humans benefit from activities that generate CO_2 without bearing any of the costs that emissions impose on others, too much CO_2 is being pumped into the atmosphere. The standard economic remedy calls for raising the price of CO_2 to give individuals an incentive to economize. The advantage of relying on the price system is that it allows individuals to decide how best to economize to suit their particular circumstance, rather than having government dictate this.

However, governments would still play a role in determining the "right" carbon price. Economists Robert Mendelsohn (Yale), Gilbert Metcalf (Tufts), and Peter Wilcoxen (Syracuse) considered the two major tools for using price incentives to reduce CO_2 emissions—a carbon tax and a carbon "cap-and-trade" system. Persuasive arguments were made in favor of and against each proposal.

The economists noted that the success of such policies generally would depend on three conditions. First, they must impose fewer costs than alternative policies (such as raising fuel economy standards). Second, they must address the uneven distributional impact of both global warming and higher carbon prices. The burden of higher prices is likely to fall disproportionately on poorer consumers. And the predicted environmental and economic damage from global warming is expected to be concentrated among the world's poorest people—which would seem to reduce the incentive of wealthier countries to reduce emissions. Third, the policy must strike the right balance between flexibility and stability, because credible long-term incentives must be in place to promote necessary (and massive) private sector investments in new technology.

Either pricing approach is expected to raise fossil fuel prices by roughly 10 percent initially and more over time. Some conference participants criticized both proposals on grounds of political and governance issues, with the carbon tax generally faring better because it is less susceptible to fraud, cheating, and the self-entrenchment of constituents.

Policy Speed Bumps

Global climate change is not a local issue, and a successful effort to reduce warming (assuming this is both possible and desirable) would require a substantial degree of international cooperation. But some conference participants leveled trenchant criticisms at the Intergovernmental Panel on

Climate Change (the IPCC, co-recipient with Al Gore of the 2007 Nobel Peace Prize) and the United Nations, raising doubt about the wisdom of relying on these global organizations.

The IPCC's views on global warming are widely portrayed as representing an objective, scientific consensus, but David Henderson (a former chief economist at the OECD, now at the University of Westminster, U.K.) demonstrated that the IPCC reporting process is flawed and lacks rigor, inclusiveness, and impartiality. He cited the IPCC's poor handling of economic issues, weak peer review process, and initial refusals to disclose data for replication.

Environmental economist Ross McKitrick (University of Guelph, Ontario) affirmed Henderson's analysis with an account of his own experience with the IPCC. He showed several examples of structural bias being built into IPCC assessment reports—including the so-called "hockey stick" graph, which misrepresented evidence by excluding the Medieval Warm Period from historical temperature records. These are troubling examples, because many international policymakers look solely to the IPCC for guidance.

Even if the IPCC findings were unbiased, relying on the U.N. or similar international organizations to manage a global emissions tax or trading program would be flirting with disaster, according to journalist Claudia Rosett. Her research on the U.N. "Oil for Food" scandals suggests it would be unwise to turn over responsibility for managing trillions of dollars of global resources to an organization that cannot even audit its own books.

If it turns out that there is only a weak link between CO_2 emissions and climate warming, then dedicating trillions of dollars toward reducing emissions will prove to be foolish. Far fewer resources would then be available to adapt to any of the changes a warming climate might bring, or to do something more radical in response to impending catastrophe. Thus, as environmental analyst Kenneth Green of the American Enterprise Institute noted, uncertainty requires that we choose policies that we will not regret if we are wrong. If we do not choose carefully, we could easily do ourselves and future generations far more harm than good.

Theological and Ethical Aspects of Climate Change

There is no shortage of global problems. With so many people suffering around the world from malnutrition, disease, and other effects of immiserating poverty, and with the many problems faced by developed nations, why has climate change become the overriding issue for so many people? And should it be?

A panel of participants discussed the theological and ethical dimensions of the global warming debate. Robert Nelson (Professor of Environmental Policy, University of Maryland) described the many theological elements of the environmental movement—from reverent descriptions of wilderness areas as "cathedrals" of nature, to near-biblical messages of fear and apocalypse. Environmentalists (and society) tend to think that anything that is "natural" is good while anything that is "unnatural" is bad. This tendency has influenced the global warming debate: If we thought that global warming were a natural phenomenon, we might be more inclined to adapt to it.

Professor Henderson spoke of "global Salvationists" and their tendency to disregard the benefits and sources of material progress. Edward Kane (Professor of Finance, Boston College) closed the conference by raising some ethical questions. If CO_2 does cause global warming, do humans still have a "right" to generate it by burning fuels? What duties does each generation have to its heirs? Do these duties justify sacrificing trillions of dollars today to future generations—who almost certainly will be wealthier? Would it be more ethical to spend the money on the poor of today?

These questions underscore the trade-offs involved with global warming. The challenge is not, "Either we believe in global warming and rescue the planet or we all die." It is largely a choice between how much action to take now and how much in the future; and a choice between preventing or adapting to it. The science is still young. Some people take uncertainty as a reason to act, others take it as a reason to proceed slowly. Future generations will have the benefit of greater scientific knowledge, better technology, and greater wealth. Yet any policy might take decades to be effective (assuming we can agree on policies that are scientifically effective, economically sensible, and politically workable). The range of views expressed at our conference suggests that the debate is far from being over, and that there is still time for much-needed rational discussion. We present these selected proceedings in hopes of adding more light than heat to the debate.

WHAT IS THE CURRENT STATE OF SCIENTIFIC KNOWLEDGE ABOUT GLOBAL CLIMATE CHANGE?

Carl Wunsch

CLIMATE change may represent the most complicated of all scientific problems. It involves processes that change, superimpose, and interact with each other, on time scales from billions of years to days, including:

- The physics and chemistry of the atmosphere and ocean;
- The physics of sea and land ice;
- All of geology and geochemistry;
- Solid earth geophysics;
- The entire history of life on Earth;
- Solar physics;
- Cosmic rays; and
- Geomagnetism.

I am very impressed by people who seem to understand all of this and also are able to tell us what will (or will not) happen! In the near term, the "history of life on Earth" has to include human beings and predictions of what they will be doing to the environment over the next decades — which then brings in all of politics, economics, sociology, population dynamics, etc. What is more complicated than understanding and predicting climate change?

Numerical models are commonly used for calculating climate change in the coupled system. Although they exist in computers, these models are recognizable as machines — and they are some of the most complicated pieces of machinery ever assembled. Few are well-understood, and we know enough about computer codes to recognize that they *always* contain errors. Determining whether those errors are significant for any particular model's purpose is a formidable problem in its own right. Furthermore, extensive experience shows that coupling simple sub-components of physical systems can lead to bizarre and unexpected behavior: Consider only what happens if one hangs a simple pendulum at the bottom of another simple pendulum. Coupling ocean, ice, atmosphere, hydrologic, and biological models, each of which has its own obscurities and uncertainties, leads science into entirely new domains involving poorly understood instruments being used to probe the climate system.

Greatly compounding the difficulties is the gross inadequacy of the observational record. A truism of science is that to understand a phenomenon, one must observe it. As will be discussed further below, the climate system has phenomena in it whose time scales range from seconds to billions of years. The longest instrumental record is a single data set of temperatures in central England. Records of the past usually are extremely indirect measurements of so-called proxies (e.g., the ratios of oxygen isotopes interpreted as atmospheric temperatures) at a handful of locations around the world. Consider what modern meteorology would be like if the only data were the surface temperatures in central Greenland and Antarctica. Even today, large parts of the system are barely observed at all. Can one make skillful predictions under such circumstances?

It is important to recognize that no one has a grasp of the entire problem of climate change. Even the most expert of scientists wears at least three hats:

(1)   True expertise in a sub-component of climate;
(2)   A working knowledge of neighboring fields; and
(3)   One's role as a citizen, parent, etc.

I will state my point of view up front: We do understand a great deal about the climate system, and we know it has been extremely different in the past. Climate response to anthropogenic forcing cannot be predicted in any conventional sense, but we know enough about elements of the *potential* response to be seriously worried about what might happen. Much of the science is less definitive than many writers and speakers would have you believe, but scientists are necessarily salesmen/women, and in addition, many are troubled by the possibility for disaster in the future.

The Range of Views

Extreme views make better, more exciting stories and seem better able to move the political system. I believe that the only rational view of climate is that we do not understand it well enough to make true forecasts, but that

8

we do understand it well enough to delineate at least some of the major threats to our well-being. (Whether we are aware of all of the possibilities seems somewhat doubtful, but the completely unexpected is difficult to discuss.)

Inevitably, it seems, much of the public discussion has been captured by story-tellers, generally worried citizens, publicity seekers, etc. I distinguish between people with a strong point of view who do not pretend to be scientists (Al Gore is the leading example) and those who are scientists but who proclaim knowledge far beyond what the science supports. Scientists are held to a different standard than ordinary citizens, even if the latter are lawyers and politicians.

We are talking about an immature science confronted with societal insistence on true prediction when none is scientifically possible, given the state of understanding. We do not know whether the system is predictable and, if so, in what elements. Understanding and predictability are very different, and even if collectively we grasped fully what is happening today and what happened in the past, there is no guarantee that we can predict what the future holds.

Absent scientific maturity, the field of climate change study gives rise to Elmer Gantry figures. It is remarkable how such a complicated problem attracts so many vociferous non-expert opinions. (One hopes that the pilot of a Boeing 747, having trouble with an engine, does not go back and poll the passengers about what his next actions should be.)

Further complications in the debate lie with the human wish for a deterministic cosmos and the human evolutionary tendency to see patterns, even where none exist, in nature. The well-described general human inability to make appropriate intuitive inferences from statistical data greatly muddies the waters. So searches for periodicities and the like take on elements of the search for the Holy Grail. Random walks are very unintuitive for most people (see, for example, the phenomenon of "gambler's ruin" and studies by economist Daniel Kahneman and others) but may in fact dominate much of the climate system. (I discuss this issue further in Wunsch, 2008, with many more references.)

Stopping the Gulf Stream

Some of the sensational flavor of the discussion is captured by news accounts such as this newspaper article, which appeared in *The Guardian*

(London) on October 27, 2006, and was written by James Randerson, science correspondent for that paper:

Sea Change: Why Global Warming Could Leave Britain Feeling the Cold
- No new ice age yet, but Gulf Stream is weakening
- Atlantic current came to halt for ten days in 2004

Scientists have uncovered more evidence for a dramatic weakening in the vast ocean current that gives Britain its relatively balmy climate by dragging warm water northwards from the tropics. The slowdown, which climate modelers have predicted will follow global warming, has been confirmed by the most detailed study yet of ocean flow in the Atlantic. Most alarmingly, the data reveal that a part of the current, which is usually 60 times more powerful than the Amazon River, came to a temporary halt during November 2004.

The nightmare scenario of a shutdown in the meridional ocean current that drives the Gulf Stream was dramatically portrayed in the movie, *The Day After Tomorrow* [2004]. The climate disaster film had Europe and North America plunged into a new ice age practically overnight.

Although no scientist thinks the switch-off could happen that quickly, scientists do agree that even a weakening of the current over a few decades would have profound climate consequences. Warm water brought to Europe's shores raises the temperature by as much as 10°C in some places, and without it the continent would be much colder and drier.

Researchers are not sure yet what to make of the ten-day hiatus. "We'd never seen anything like that before and we don't understand it. We didn't know it could happen," said Harry Bryden, at the National Oceanography Center, in Southampton, who presented the findings to a conference in Birmingham [U.K.] on rapid climate change.

Is it the first sign that the current is stuttering to a halt? "I want to know more before I say that," Professor Bryden said.

Lloyd Keigwin, a scientist at the Woods Hole Oceanographic Institution, in Massachusetts . . . described the temporary shutdown as "the most abrupt change in the whole [climate] record."

He added: "It only lasted ten days. But suppose it lasted 30 or 60 days; when do you ring up the prime minister and say, 'Let's start stockpiling fuel? How can we rule out a longer one next year?'"

Professor Bryden's group stunned climate researchers last year with data suggesting that the flow rate of the Atlantic circulation had dropped by about 6,000

10

tons of water a second from 1957 to 1998. If the current remained that weak, he predicted, it would lead to a 1°C drop in the U.K. in the next decade. A complete shutdown would lead to a 4°C-6°C cooling over 20 years.

The study prompted the U.K.'s Natural Environment Research Council to set up an array of 16 submerged stations spread across the Atlantic, from Florida to North Africa, to measure flow rate and other variables at different depths. Data from these stations confirmed the slowdown in 1998 was not a "freak observation"— although the current does seem to have picked up slightly since.

There also were special reports on this notional event in the scientific periodicals. (*See,* e.g., Bryden et al., 2005; D. Quadfasel, 2005.)

These extreme views are matched on the other "side" by the so-called "skeptics," although a much better term would be "deniers," because all good scientists are skeptics.

Global Warming Deniers

Here are some excerpts from an essay by retired physicist Fred Singer (Hoover Institute website, 2000) amplifying his congressional testimony:

The essay [Singer's] also demonstrates that global warming (GW), if it were to take place, is generally beneficial for the following reasons:

1. One of the most feared consequences of global warming is a rise in sea level that could flood low-lying areas and damage the economy of coastal nations. But the actual evidence suggests just the opposite: A modest warming will reduce somewhat the steady rise of sea level, which has been ongoing since the end of the last Ice Age—and will continue no matter what we do as long as the millennia-old melting of Antarctic ice continues.

2. A detailed reevaluation of the impact of climate warming on the national economy was published in 1999 by a prestigious group of specialists, led by a Yale University resource economist. They conclude that agriculture and timber resources would benefit greatly from a warmer climate and higher levels of carbon dioxide and would not be negatively affected, as had previously been thought. Contrary to the general wisdom expressed in the IPCC report, higher CO_2 levels and temperatures would increase the GNP of the United States and put more money in the pockets of the average family.

There are extreme views on both sides. Scientists predicting imminent "shutdown" of the North Atlantic Ocean circulation are talking scientific nonsense. It is somewhat forgivable if a non-scientist like Al Gore quotes them, but working scientists are supposed to be able to distinguish what they know from what they regard as possible and fear. Singer's dogmatism—he cannot possibly be sure of his assertions—is equally ill-founded, yet he claims credentials as a scientist that Gore does not.

Do We Know Anything for Certain?

In general, scientific understanding and certainty exist on a continuum of levels of confidence. One could ask an individual scientist to lay a large bet and then see what kind of odds he/she would give. For me, on a confidence scale of 1 to 10, I list the following:

Cases with a great deal of certainty where I would be willing to bet an unlimited amount of money:

1: (Almost) Absolute Certainty:
   The speed of light is an upper bound on finite mass velocity (do tachyons exist?).
   The second law of thermodynamics always holds.
1+: The Gulf Stream cannot turn off (unless the sun goes out or the Earth stops rotating) owing to the principle of conservation of angular momentum.
2: I would give very large odds:
   The sun will rise tomorrow.
   The vernal equinox will occur on 20 March at 6 a.m. in 2008. (I could go out about 10 million years, but I cannot go out 30 million years because the limit of orbital predictability lies somewhere in between.)
3: Next June will be warmer in the Berkshires than this February (but recall 1816, "the year without a summer").

Cases in which confidence, and so a bet, would be circumstance dependent:

4: It won't rain tomorrow. (Depends on the general meteorological situation.)

Cases in which confidence is no better than chance:

9+: It will rain on this date one year from today (details depend on clima-
    tology).

In a climate context, I have the following degrees of confidence:

2: Atmospheric CO_2 has been rising for decades and the rise is primarily
   anthropogenic in origin.

Increasing atmospheric CO_2, under present conditions and on time scales
of decades to hundreds of years, will:

2: Raise the global average temperature;
2+: Make the ocean more acid;
3: Accelerate sea level rise;
3: Warm the poles more than the tropics;
3: Melt the tundra;
5: Warm nights more than days;
?: Make hurricanes more intense/more plentiful;
?: Make the U.S. Midwest become much drier; and
8-9: Make the Massachusetts Berkshires become wetter.

So the climate problem involves the full range of scientific uncertainty,
and it is very difficult to communicate the different levels of understanding
and predictability in two-minute sound bites.

Why Is the Scientific Uncertainty about Climate So Variable?

As is already noted, the instrumental records begin only in the 1600s.
The central England annual mean temperature (Manley, 1974) is probably
the longest existing instrumental record of climate in existence. Compared
to the known time scales of climate change, the record is extremely short,
and no other such record exists from anywhere else.

Atmospheric data only come to reasonably represent the whole northern
hemisphere in the late 19th century. The biggest improvement was after
World War II, when a semblance of global coverage began. Oceanic data

did not become even quasi-global until about 1992.

More generally, consider another, emblematic record that purports to show the temperatures over tens of thousands of years as seen in a central Greenland ice core sample. The record shows a very interesting mixture of high and low frequencies, apparent abrupt jumps, at least one plateau, and up/down asymmetry.

Another ice core record is similar, except that it comes from Antarctica (see IPCC, 2007), it covers a period of over 400,000 years, and it shows not only estimated temperature but also the carbon dioxide content of the atmosphere and the atmospheric dust content. Some scientists interpret this record as showing that carbon dioxide changes caused the glacial/interglacial cycles. Others state that the temperature changes caused the carbon dioxide shifts and, more likely, that there was a complicated interaction between the two, not reducible to a simple statement of cause and effect.

Consider the length of the instrumental record, covering central England since 1650, relative to the huge variety of much longer time-scale phenomena represented in the two ice core records.

These and similar records represent the superposition, and interaction, of extremely diverse phenomena taking place on time scales of change from a decade to 100,000+ years. The physics varies with frequency, but looking at the data, one's eye tends to see events—some of which *may* be random superposition of fluctuations taking place over a broad range of time scales. Change is always occurring, and no single sentence can summarize the changes, nor their physics, chemistry, and biology.

We can conjecture lots of somewhat plausible hypotheses from these observations, but none is provable. CO_2 goes up and down roughly with temperature, but there is no simple time lag (they interact, but they interact *differently* on different time scales), and so there is no simple causal story. The dust record, just by way of example, is sometimes interpreted as meaning that glacial periods were drier (and therefore dusty), or windier (and therefore more dust was transported to Antarctica), or both. This ambiguity is typical of the use of proxy records.

To physicist Niels Bohr (and many others) is attributed the saying, "Prediction is very difficult, especially about the future." And so it is.

Several climate models (*see* Seager et al., 2007) of net precipitation in the United States Southwest between 1880 and 2100 provide another example of this. They not only differ in the historical period but, as predictions, the results range from severe drought to little change. Deniers justifiably can point at such results and assert that there is no apparent predictive skill in the models. On the other hand, the possibility of drought seems very real (the particular extreme model is as credible as the others), and my own inference is that we must be aware that such an outcome is possible — but not guaranteed.

Another example involves various model estimates of the September ice cover between now and 2100. The results (Eisenman et al., 2007) range from near 0 to near 100 percent. With this range of model outcomes, as an indicator of skill, should one be surprised by the sudden reduction in ice cover observed last summer (2007) as reported by A. Revkin, "Arctic Melt Unnerves the Experts," *New York Times*, October 2, 2007? It was not predicted, but given the range in the models, one might argue that they reflect a general failure to capture the full physics of the system. The elements controlling sea ice cover are insufficiently observed and understood to be making true predictions running 100 years into the future. That, of course, does not mean that nothing dire will happen!

How should one regard models? Numerical models of the climate system are extremely complicated machines that are attempts by very serious scientists to encompass many of the most important elements of the climate system. Despite huge efforts, there is little or no evidence that they are sufficiently realistic when run over decades to hundreds of years to be credible. They are credible as sketches of conceivable future climate states. They might be regarded as analogues of the novel — fictional — but the best of them illuminate important components of the climate system, without being taken literally. *Crime and Punishment* and *Madame Bovary* are very important books discussing human nature; no one confuses them with literal history. A really good climate model is similarly neither history nor prognostication — it casts light on elements of the problems of climate change.

Prediction is hard!

A Case Study: Modern Sea Level Rise

In his remarks (directed at Congress), Singer suggests that sea level rise is well understood, that there is nothing we can do about it, and that it will simply continue as it has for thousands of years. I will focus briefly on this climate sub-problem because I claim to know a bit about it (*see* Wunsch et al., 2007).

The general background is this: Sea level has been rising for about 20,000 years, having nearly stabilized (but not completely so) about 8,000 years ago. The "worry" begins with graphs (e.g., from Church et al., 2001) showing IPCC scenarios of future sea level rise, whose social consequences are quite alarming. I will not discuss the societal impact but will only ask how much we understand of what is happening today, so that we might evaluate the scenarios.

The effects of sea level rise are already visible in some places. An aerial view of marshes near Cambridge, Maryland shows progressive drowning and loss of coastal marshes during the period from 1938 to 1988, as a result of rising sea level (Douglas et al., 2001).

Satellite altimetry, whose record begins in 1992, is widely believed to be the best available estimate of global sea level rise in the last 15 years (noting, again, how short that interval is). The global pattern of rise is, however, extremely complicated, with large regions of falling sea level over that interval. Serious questions arise as to the accuracy of a global average in the presence of this complexity (and the inevitable errors existing in the measurement).

The rate of rise has been much greater in the past, including changes that look like major jumps. So why all the fuss (echoing the "skeptics")? There are several answers to this question. First, geologically, the present pace of CO_2 increase is unprecedented, and rates of change have very powerful implications for system response (doubling CO_2 over 10,000 years will produce a very different result than doubling it over 100 years). Second, and at least as important, there are now over six billion people in the world — a totally unprecedented situation — indeed, one can argue that climate change is an issue primarily because of human population growth.

Making inferences about sea level change prior to the advent of high accuracy altimetry and making independent confirming measurements during the altimetric era (altimetry is a remarkably complicated system) proves very difficult for a number of reasons. There are many questions, too, about the technologies used in the past and whether they give spurious trends.

The threats of future abrupt change concern primarily the possibility that the existing continental ice sheets in fact might be partially unstable (Weertman, 1974; Bentley, 1997), particularly the West Antarctic ice sheet. It proves very difficult to determine how much of the present rate of rise is due to melting ice, and the dynamics of continental glaciers are poorly understood.

As with so many elements of climate change, a significant scientific body believes that it can provide analyses far beyond what the observations or models seem to support. Those scientists are not necessarily wrong; it is just that the science does not support a definitive inference.

Despite all these difficulties, some people persist in publishing papers showing sea level in the past (and future), as though the statistical error in these predictions—the error bars in their charts—were negligible, and drawing strong inferences, such as a correlation with global mean temperatures, which are themselves subject to major uncertainties. Can one reconstruct global sea level and/or temperature back to 1875? Science can be brought into disrepute this way.

As discussed at length by Wunsch et al. (2007), the determination of the rate of rise of modern global sea level lies at the very edge of what is technically possible. Great care is required in discussing what is known or not known.

My Own Inferences

Sea level is almost surely rising, inexorably, at a global average rate somewhere between 1 and 3 millimeters per year. This is slow physics, probably around category 2 in the list of scientific certainty above. Acceleration of sea level rise over the next several decades is a plausible possibility, maybe a 4 on the certainty scale.

An abrupt event (e.g., breakup of the West Antarctic ice sheet) is physically possible, but its probability is beyond our ability to calculate with any confidence—but that is not the same as saying that it is not going to happen! We are unlikely to know a great deal more about sea level rise 10 or 20 years from now because of the slow time scale problem. One hundred years from now might be a different story.

What should we, as scientists, say to the public about all this? What should we, as citizens, ask our governments to do? What should we do as individuals?

Al Gore is worried. He does not get all of the science right (but few scientists do either—the problem is beyond any one person's ability to encompass all elements). Some of the public discussion makes me cringe. But I am worried too. I think we need to take precautions.

Summary and Conclusions

Climate change is an extremely difficult problem to reconstruct, understand, and forecast. Unlike many difficult scientific problems, it has immediate social, political, and economic consequences. In the presence of societal demands for predictions of what the future holds, an immature science holds few definitive answers and opens the door to dogmatism, charlatanism, exaggeration, and, ultimately, distortion of the science.

The science is not able to provide definitive predictions even of such comparatively simple elements as the future rate of global sea level rise. The science *is* mature enough to describe credible threats to the future environment and human well-being. Decisions about precautionary steps must be made in the presence of the scientific uncertainty, an uncertainty that is going to persist for decades, perhaps indefinitely.

Good scientists are always open to the possibility that they are wrong. I hope I, and others who share my worries, are wrong. I wonder if Fred Singer ever asks himself if there is any possibility that he could be mistaken. Science is *always* a working consensus, ordinarily because it has predictive power. Pre-plate tectonics geology is now known to be conceptually wrong in a fundamental way, but it was extremely useful, nonetheless, in finding oil and ore deposits and in outlining the history of life.

One can fault many scientists working on climate change for overstating the justified degree of confidence in their inferences. There is a human need to tell stories, to regard the world as deterministic, and many of the scientists are truly troubled about the possibilities they see for the future climate (the parental role I mentioned above).

Science is always about the weight of the evidence. Few things are provable in an absolute sense. One must always be open to surprises, reinterpretations, and the possibility that one has been wrong.

Here are my worries based on the weight of evidence that I know:

- Continued sea level rise, with possible acceleration in its rate;
- Extended droughts/floods, with greater weather extremes generally; and
- Oceanic acidification.

In any essay on climate change, it seems unreasonable to simply ignore the largest element of risk—the one presented by human population growth. Had human population stabilized at about three billion people (about one-half of what we have today), the Earth would be in a much more resilient state. The prospects for growth to nine billion and beyond are truly discouraging (we probably are well above six billion today). Neglect of this problem borders on the criminal.

Climate change presents society with what is best regarded as an insurance problem. Like the much more familiar insurance problems we confront as individuals, there is no simple formula for what one should do. How much life insurance does someone need? Is fire insurance on one's house worth paying the premium for, year after year? Is it worth paying a large amount of money, in a precautionary sense, to rewire a house if it somewhat reduces the risk of a fire? There are no simple formulas for what an individual should do—it is a complex weighing of comfort with risk, financial resources, and intuitive calculation of odds. In general, everyone comes to a different conclusion.

Climate change produces an insurance problem on a societal scale, orders of magnitude more complicated than what individuals confront. What is so depressing is the reduction of what needs to be a continuing, informed discussion of the tradeoffs of risk, the possibility of precautionary steps, the uncertain science, etc., to political rants. To end as I began, climate change presents us with one of the most difficult of all scientific problems, heavily encumbered with the need to make political and economic decisions long before hard evidence is at hand.

References

Bard, E., B. Hamelin, M. Arnold, L. Montaggioni, G. Cabiouch, G. Faure, and F. Rougerie. "Deglacial sea-level record from Tahiti corals and the timing of global meltwater discharge." *Nature*, vol. 382 (1996), pp. 241-244.

Bentley, C. R. "Rapid sea-level rise soon from West Antarctic ice-sheet collapse?" *Science*, vol. 275 (1997), pp. 1077-1078.

Bryden, H. L., H. R. Longworth, and S. A. Cunningham. "Slowing of the Atlantic meridional overturning circulation at 25N." *Nature*, vol. 438 (2005), doi:10.1038/nature04385, pp. 655-657.

Cazenave, A. and R. S. Nerem. "Present-day sea level change: observations and causes." *Revs. Geophys.*, vol. 42 (2004), doi:8755-1209/04/2003RG00139.

Church, J., J. M. Gregory, P. Huybrechts, M. Kuhn, K. Lambeck, M. T. Nhuan, D. Qin, and P. L. Woodworth. "Changes in sea level," in J.T.

Houghton et al., eds. *Climate Change 2001: The Scientific Basis*, pp. 639-693. New York, NY: Cambridge University Press, 2001.

Douglas, B. C., M. S. Kearney, and S. R. Leatherman, eds. *Sea Level Rise: History and Consequences*. San Diego, CA: Academic Press, 2001.

Eisenman, I., N. Untersteiner, and J. S. Wettlaufer. "On the reliability of simulated Arctic sea ice in global climate models." *Geophys. Res. Letts.*, vol. 34, no. 10 (2007), L10501.

Intergovernmental Panel on Climate Change (IPCC). *Climate Change 2007—The Physical Science Basis*. Cambridge, UK: Cambridge University Press, 2007.

Manley, G. "Central England temperatures: monthly means, 1659-1973." *Quat. J. Roy. Met. Soc.*, vol. 10 (1974), pp. 389-405.

Quadfasel, D. "Oceanography: The Atlantic conveyor slows." *Nature*, vol. 438 (2005), pp. 565-566.

Rahmstorf, S. and 6 others. "Recent climate observations compared to projections." *Science*, vol. 316 (2007), p. 709.

Rignot, E. and P. Kanagaratnam. "Changes in the velocity structure of the Greenland ice sheet." *Science*, vol. 311 (2006), pp. 986-990.

R. Seager and 12 others. "Model projections of an imminent transition to a more arid climate in southwestern North America." *Science*, vol. 316 (2007), pp. 1181-1184.

Weertman, J. "Stability of the junction of an ice sheet and an ice shelf." *J. Glaciol.*, vol. 13 (1974), pp. 3-11.

Wunsch, C., R. Ponte and C. Wunsch. "Decadal trends in sea level patterns: 1993-2004." *J. Clim.*, vol. 20 (2007), pp. 5889-5991.

Wunsch, C. "Extremes, patterns, and other structures in oceanographic and climate records," in C. Garrett and P. Muller, eds. *Extreme Events: Proceedings, Aha Hulika'a Hawaiian Winter Workshop, January 2007*. Honolulu, HI: SOEST Spec. Pub., University of Hawaii, 2008 (in press).

GLOBAL WARMING: WHERE HAVE WE BEEN AND WHERE ARE WE GOING?*

David S. Chapman

L IFE has existed on planet Earth for approximately 4 billion years. In that time, climate has swung between ice ages and warm periods. But generally, Earth's atmosphere has been in chemical balance, its composition changing slowly in response to changes in geology and life on the planet's surface. Now, growing population and the by-products of civilization are upsetting this balance. For the first time in the planet's history, humans are significant agents of global change.

I am both a geophysical researcher and an educator, teaching courses on global environmental change. Geoscientists largely study the past. But I am also a grandparent of four and am concerned about what the world is going to look like for at least their lifetimes throughout the 21st century. Thus my presentation uses the eyes of a geoscientist to look back in time to see where we have been but also uses strong inferences to predict where we might be going with respect to global warming.

Quantitative Evidence of Global Warming Trends

We have quantitative evidence from three independent sets of measurements that widespread warming of the Earth's surface has occurred in the last century. The data sets are surface air temperature measurements, sea level changes, and temperature profiles in boreholes.

Direct temperature measurements at weather stations suggest that the surface of Earth has warmed, on average, about 0.9° C (1.6° F) in the last 100 years. The evidence comes from temperatures routinely measured and reported daily for thousands of weather stations around the world, both on land and at sea. Daily temperatures are combined to produce average weekly, monthly, and annual temperatures. We can track how the average annual temperature changes from year to year.

A summary picture of all those temperature measurements for the last 150

---

* This article is adapted from Prof. Chapman's talk at the conference. Due to production constraints, it excludes some of the charts, photographs, and illustrations that he included in his presentation. – Editor.

years is shown in Figure 1. Measurements have been combined to produce an average temperature for the entire globe each year. It is important to explain what Figure 1 is and what it is not. Precisely, it is a graph of the average temperature at Earth's surface on land and at sea. On land, air temperatures are measured at weather stations by thermometers mounted 1 to 2 meters above the ground surface. At sea, the temperature of the sea surface water is measured along ship tracks. Much effort has been spent in correcting raw measurements to compensate for changing technology over time, such as the change from mercury thermometers to digital electronic thermometers and the change in locations of weather stations. Statistical methods have been used to close gaps in the record. The effect of large population centers, called the "urban heat island effect," also has been calculated, but it accounts for less than 15 percent of this century's global warming.

Between 1860 and 1910, the global mean surface temperature fluctuated between 13.5° and 13.7° C with no obvious trend. Temperatures rose

Figure 1. Global warming revealed. Air temperature measured at weather stations on continents and sea temperature measured along ship tracks on the oceans are combined for a global mean temperature each year. This 150-year time series is the direct instrumental record of global warming. The curve is smoothed with a five-year running average.

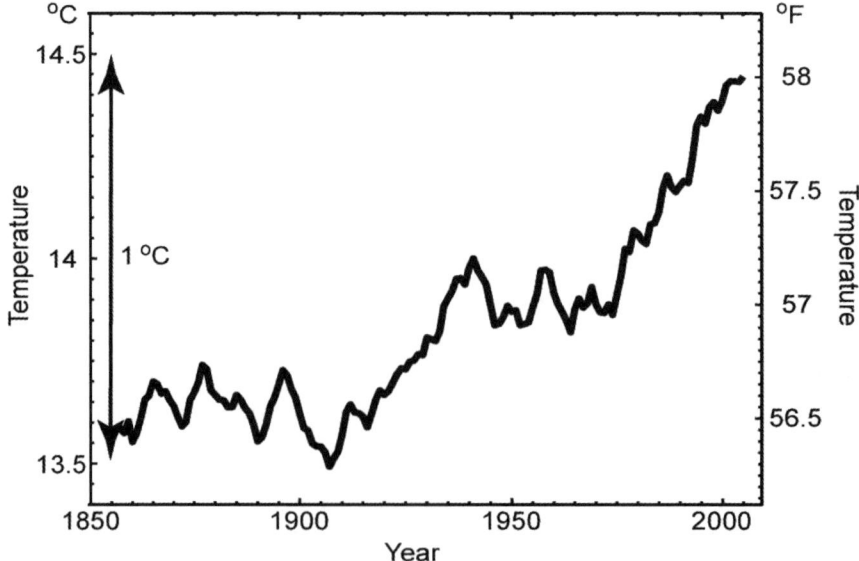

rapidly between 1910 and 1945, stabilized for three decades, and then rose again dramatically after 1975. Current global mean temperatures are significantly warmer than any previous decade in the last century. The global mean temperature has increased by 0.1° C per decade for the last 20 years, with 2005 being among the warmest years on record.

Not all parts of the globe exhibit the same amount of warming. Low-latitude regions generally have warmed less; high-latitude regions, more. Some areas even have exhibited slight cooling over the same time period. Furthermore, warming is not uniform in time. All regions of the globe have experienced years of cooler temperatures embedded within the warming trend. The irregularity of warming in both time and space merely indicates a chaotic component in climate change.

A confirmation of this century's warming trend comes from a completely independent set of measurements. Sea level is rising (Figure 2). The volume

Figure 2. Sea level is changing. Observing stations from around the world report year-to-year changes in sea level. The reports are combined to produce a global average time series. The year 1990 is arbitrarily chosen as zero for display purposes.

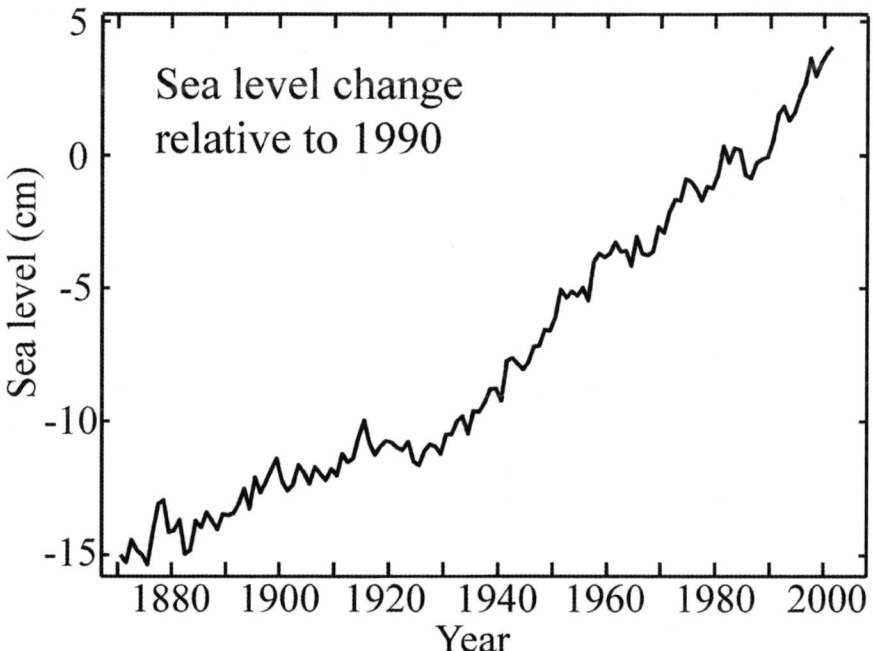

of water in oceans is increasing because glaciers and ice caps are melting and because water already in the oceans is expanding as it is being heated. As with temperature observations, sea level is measured at many sites each day. Daily fluctuations, caused principally by tides and storms, are averaged out, and the slow changes over time are charted. Mean sea level has risen by nearly 20 centimeters since 1880. Currently, it is rising at the rate of more than 3 millimeters per year.

Sea level changes, like global temperature changes, are not steady. Neither are the detailed changes entirely synchronous with surface temperatures. Thermal expansion of the water column tends to come later than the corresponding change in surface temperature, and the detailed differences are highly affected by ocean currents. The irregularity in sea level rise is another indication of the complex natural world that we live in.

A third confirmation that Earth's surface has warmed in the last century comes from an unlikely source, the solid ground below our feet. Temperature of rock in the outermost part of the Earth records a history of climates of previous centuries (Figure 3). This archive of past temperatures is accessed simply by lowering a sensitive thermometer in a borehole to obtain a profile of temperature with depth below the surface. To isolate a climate signal within the profiles, we must make corrections for the expected steady temperature increase caused by heat flowing out of Earth's interior and for the oscillations caused by daily and seasonal temperature changes at the surface.

The geothermal method of measuring the Earth's temperature yields two kinds of information. First, the difference of temperature at the surface, compared to the extrapolation of the deeper or older temperature, shows how much the area has warmed. Second, the depth at which a warming trend emerges in the temperature-depth profile, just through the calculations of thermal physics, tells us how long ago that warming started. Rock properties govern how fast temperature changes propagate into the subsurface. Temperature changes that started 100 years ago are visible in the temperature-depth profile to a depth of about 100 meters. All changes in climate that have occurred in the past millennium theoretically are imprinted in the uppermost 500 meters of the crust, a depth easily attainable by inexpensive drilling.

Already several geothermal data sets from North America have been analyzed for evidence of surface temperature changes (Figure 3). In all our datasets, temperatures are warmer near the surface than they are in the subsurface in the reduced, transformed temperature field. *Investigations in the Alaskan Arctic* by Arthur Lachenbruch and his colleagues at the United States Geological Survey provided dramatic evidence of warming. Temperature profiles from wells spread across 500 kilometers of northern Alaska show anomalous warming of 2° to 5° C in the upper 100 to 150 meters of the permafrost and rock. The depth where warming is evident shows that surface climate change in Alaska had a 20th century onset. Borehole temperature profiles in Eastern Canada document a less dramatic but equally clear warming of more than 1.0° C. Nebraska and Utah sites both exhibit warming of about 0.5° to 1.0° C, also starting around the turn of this century. The geographic variation in warming seen in weather station data, greater at high latitudes and lesser at low latitudes, is mimicked in

Figure 3. Borehole temperatures confirm widespread warming in North America. Composite borehole temperature profiles from groups of sites in North America show diagnostic hooks toward warmer temperatures within the uppermost 100-150 meters near the surface. The profiles suggest substantial warming in the last century from 0.6° C in southeast Utah to more than 2.0° C in Alaska. Curves are offset arbitrarily for display purposes.

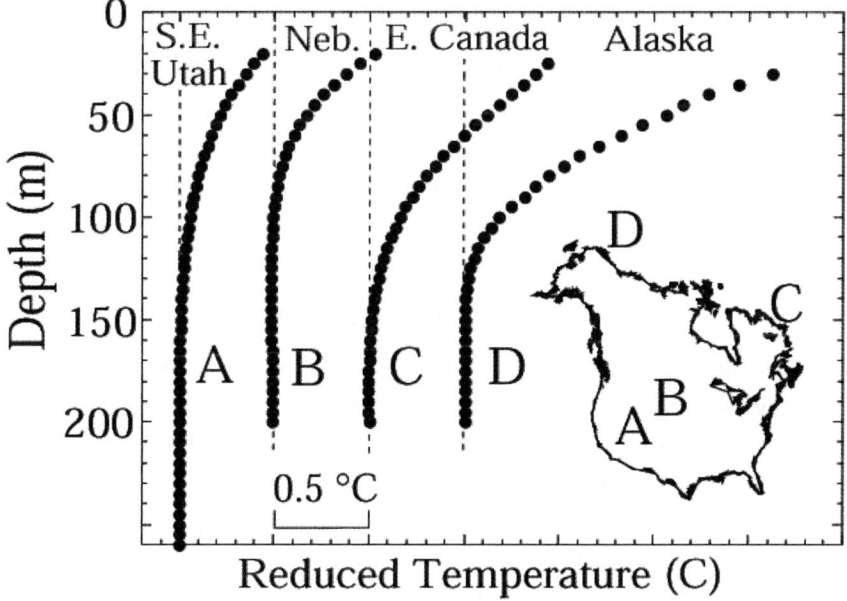

the geothermal data.

The geothermal analysis provides more information than a simple confirmation of 20th century warming. For each site it is possible to infer an 18th or 19th century baseline temperature that existed prior to any instrumental record. This is a critical period of time that connects the start of the industrial revolution to the present century.

Other Evidence of Global Warming

There are many spectacular pictures that provide additional, albeit primarily qualitative, evidence for unusual warming on planet Earth. For example, the melt area of the surface ice in Greenland, which dominates land ice in the Arctic region, increased by about 20 percent from 1979 to 2005. Recent satellite observations suggest that ice accumulation on the Greenland ice sheet dome is now less than losses due to melting around the edges.

Another example is coastal erosion in Alaska, a consequence of less ice due to melting in spring and fall. Professor Wunsch mentioned that he was not surprised that there was less Arctic ice; but less ice has consequences. On the coast of the Bering Sea, the Arctic ice acts as a breakwater, preventing big waves from breaking against the shore. The village of Shishmarev recently lost 125 feet of shore to coastal erosion in one storm. In this way, a group of people who probably do not consume much energy or put much CO_2 into the air are victims—passive victims.

Warming also has economic consequences. According to data from a recent publication in *The Arctic Assessment*, the number of travel days in the North Slope of Alaska from 1970 to the present has fallen from about 200 days to about 120 days. Land travel is possible only when the ground is frozen, and there have been progressively fewer days of frozen ground over this period.

Causes of Global Warming

What controls planetary surface temperatures and their changes through time? We know that an energy balance determines Earth's surface temperature. Incoming solar radiation is balanced by outgoing thermal radiation from the planet. Incoming solar radiation depends on the solar output and the distance of the planet from the sun and is independent of Earth's surface temperature. Outgoing thermal radiation, on the other hand, has a very strong dependency on Earth's surface temperature. If planets were born with a cold surface, they would warm quickly because the incoming radiation would exceed outgoing radiation. Surface temperature, and thus outgoing radiation, would increase until a balance was reached. On the other hand, planets born hot soon would have surfaces that were relatively cool with respect to their cores because of the high rate of heat loss due to radiation. In a relatively short time, the surface temperature would depend more on the radiation equilibrium than on the internal core temperature of the planet.

A planetary atmosphere serves two functions. Clouds reflect incoming radiation back to space, providing a lower radiation equilibrium temperature than if there were no clouds. Atmospheres, like Earth's, also contain greenhouse gases such as water vapor, carbon dioxide, and methane. Adding greenhouse gases in their natural abundance traps some of the outgoing thermal radiation and reestablishes a new energy balance with a warmer temperature. This warming is referred to as the natural or beneficial greenhouse effect. The combination of the strength of our sun, the distance of the Earth from the sun, reflective properties of our clouds, and the natural greenhouse effect have produced a surface temperature of +15° C.

Satellite observations of "earthlight" confirm details of the greenhouse effect for Earth (Figure 4). Earthlight is the outgoing thermal radiation from Earth. Without an atmosphere to absorb radiation, the radiance would be mapped by a smooth curve called the spectrum, which peaks at a wavelength between 15 and 20 micrometers, nearly 40 times longer than the wavelength of incoming visible light. But not all of the radiation gets out. The unshaded regions of Figure 4 show how much radiation is absorbed by the greenhouse gases water vapor (45 percent), carbon dioxide (30 percent), and methane (20 percent). Other minor greenhouse gases account for the remaining absorbed radiation.

Although most of the natural greenhouse effect is due to water vapor, the amount of water in the atmosphere is changing relatively little because of human activities. The anthropogenic greenhouse gases are carbon dioxide, methane, nitrous oxides, ozone, and chlorofluorocarbons (CFCs). Of these, the most important human-derived greenhouse gas is carbon dioxide. The addition of greenhouse gases into the atmosphere leads to a net increase in radiative forcing. In essence, this means that human activity has upset the energy balance of our planet, resulting in warming.

The focus on CO_2 occurs because of its central place in this diagram (Figure 5). We know that atmospheric CO_2 concentrations for the past 300 years were more or less level until around 1750 and have increased steadily since then, with recent increasing rates of change. The historic (pre-1750) level was about 280 parts per million; now we are at 385 parts per million. Each year, the burning of fossil fuels adds about 5.4 billion tons of carbon to the atmosphere. Deforestation accounts for another 1.6 billion tons by reducing the storage of carbon by trees. The result is that the atmospheric level of carbon dioxide has increased 30 percent since 1750 (Figure 5). The carbon dioxide buildup in the atmosphere has been measured instrumentally since 1959 at an observatory near the summit of Mauna Loa in Hawaii. Each year, atmospheric carbon dioxide rises and falls by almost 6 ppmv, signaling the growing and dormant seasons for plants (Figure 5, inset). But each year the annual maximum and minimum increases by about 1.5 ppmv (parts per million by volume). This growth of carbon dioxide concentration

Figure 4. Earthlight confirms greenhouse effect. Thermal radiation emitted from the Earth's surface and atmosphere is observed from a satellite instrument looking down at Earth (irregular line). In the absence of a greenhouse effect, the radiance would follow the solid smooth curve. The difference between the smooth and irregular lines represents the radiation absorbed by greenhouse gases in diagnostic wavelength bands. Units of radiance are watts per square meter per steradian per wavenumber. (After Houghton, 1997.)

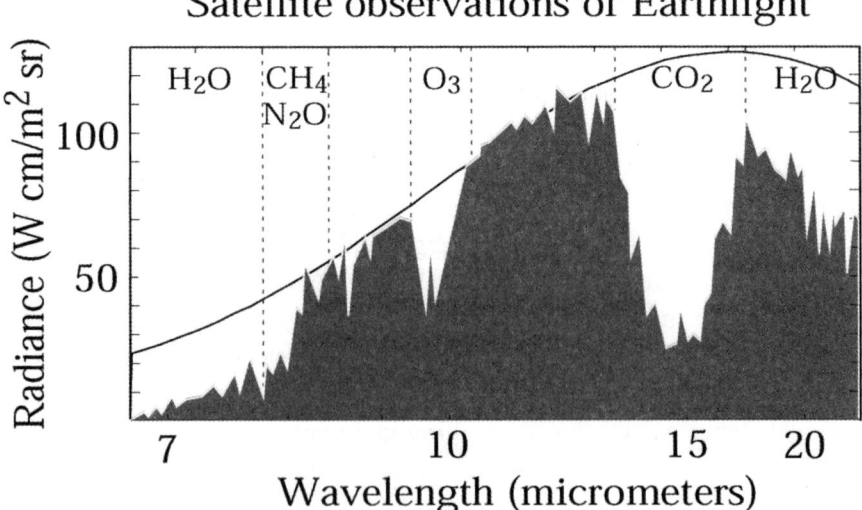

Satellite observations of Earthlight

is particularly problematic because the gas has an average lifetime of 100 to 150 years in the atmosphere, being depleted mainly by dissolution in the oceans over time. If we were to decrease our input of carbon dioxide dramatically today, the effects of the present concentration of carbon dioxide would still be felt for another 150 years.

We should not focus entirely on the critical role of CO_2 as a greenhouse gas without examining other potential causes of surface temperature change. Figure 6 gives the results of the most recent IPCC assessment for factors that have contributed to radiative forcing (RF) change between 1750 and 2005. Seven factors are considered: long-lived greenhouse gases, ozone, stratospheric water vapor from methane, surface albedo (reflection), aerosol effects, contrails, and solar irradience changes. Aerosols and surface albedo changes have produced a negative radiative forcing, and thus cooling effect,

Figure 5. Rise in atmospheric carbon dioxide. The concentration of carbon dioxide in Earth's atmosphere has increased steadily from 280 to 385 ppm since 1700. Early data come from gas bubbles trapped in ice. Since 1959, carbon dioxide concentration has been measured at observatories in Hawaii and elsewhere. Because the atmosphere is well mixed, the ice core data blend seamlessly into the Hawaiian instrumental record. Recent measurements show uptake and release of carbon dioxide with seasons (inset) superimposed on the steady global increase.

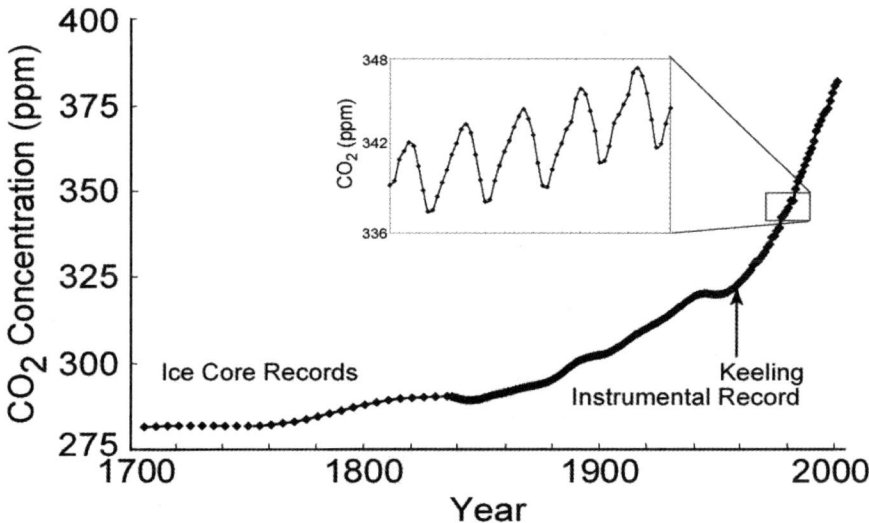

since 1750. But positive radiative forcings (warming) exceed the negative radiative forcings (cooling), and the strongest forcing is from CO_2. Solar irradience also contributes a positive forcing (warming) over this time period, but the net effect is small. The level of scientific understanding (LOSU column in Figure 6) of most forcings still needs improvement.

Looking Back in Time

So far we have examined greenhouse gases, global warming, sea level, and borehole temperatures in the time frame of the last 300 years. That is too short a time span, however, to reveal natural variations inherent in Earth processes. Fortunately, much longer time series of atmospheric chemistry and temperature, as long as 650,000 years, are available by examining air

Figure 6. Radiative forcing (RF) for the Earth's climate system, 1750 to 2005. Factors may cool (negative forcing) or warm (positive forcing) the planet. The bars in the graph are the best estimate of how much these factors have warmed or cooled the Earth, and the little pincer lines show the amount of uncertainty. The estimated net anthropogenic amount is 1.6 W/m² (watts per square meter). LOSU in the rightmost column of the graph is an abbreviation for "level of scientific understanding," as estimated by the IPCC.

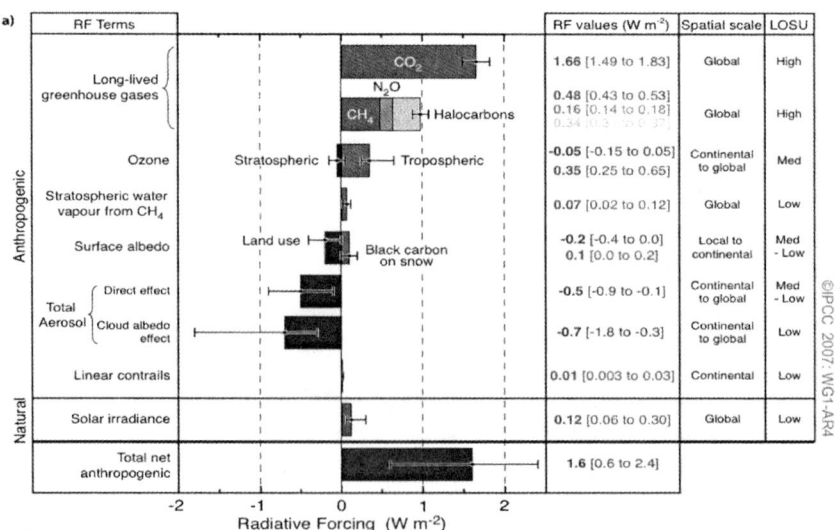

bubbles trapped in ice and the chemistry of the ice itself from both Greenland and Antarctica. We can ask how much CO_2 changes without human involvement and how much of the CO_2 greenhouse effect might be natural or unnatural.

Figure 7 shows an Antarctic ice core record for the last 165,000 years and demonstrates convincingly that atmospheric CO_2 does change without human involvement. Over this time period there is a regular pattern between low concentrations of 200 ppm and high concentrations of about 300 ppm. The time span of the cycle is about 100,000 years, with smaller oscillations of about 20,000 and 40,000 years within the larger pattern. These three periods are known as Milankovitch cycles, resulting from the collective

Figure 7. Antarctic ice core records of CO_2 concentration and temperature for the last 165,000 years. Temperature and CO_2 change in a very similar pattern.

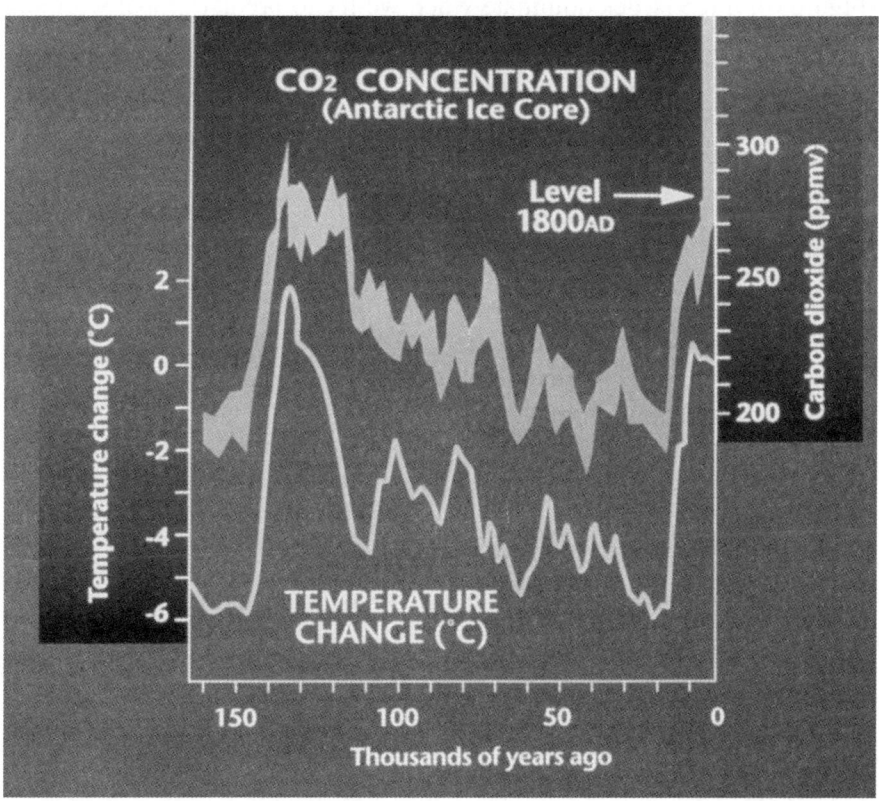

changes in eccentricity, axial tilt, and precession of the Earth's orbit. The Milankovitch cycles are thought to cause the glacial cycles, but the radiative forcing is small and so must be aided by feedback mechanisms such as CO_2 drawn into and released from the oceans in a synchronous pattern. Figure 7 shows dramatically that temperature changes mimic atmospheric CO_2 changes, strengthening the argument for greenhouse forcing. Repetitive cycles of the pattern similar to Figure 7 are now known for Antarctica for the last 650,000 years.

The unnatural part of Figure 7 is the rapid change of atmospheric CO_2 from 280 to 385 ppm in just the last 250 years. We are now at an atmospheric concentration well above any natural variation for the last 650,000 years and changing at an unprecedented rate.

The Next Century and Human Population Growth

Because of the importance of greenhouse gases for climate forcing (Figure 6), three issues dominate when we try to predict global warming scenarios for the remainder of the 21st century. Those issues are: population growth, energy consumption related to standard of living, and energy production.

Human population is now approaching 6.7 billion (Figure 8). In 1780 the population was 1 billion. Just 150 years later, in 1930, the population had doubled to 2 billion. The last doubling, from 3 to 6 billion, took only 45 years. It now takes only 11 years to add a billion people to the Earth's population, although there are indications that the growth rate is slowing. But because of the age distribution of many countries, population momentum will carry Earth's human population with certainty into the range of 9 to 12 billion. The inset of Figure 7 illustrates the predicament. Percentages of the country's population are shown for each 5-year age group for two countries. In rapid-growth countries such as Kenya, about 20 percent of the population is between ages 0 and 5. Nearly half the population is in the age range 0 to 15. It is the childbearing potential of this section of the population that propels population momentum and results in growth projections for the globe. In slow-growth countries like the United States, only 7 percent of the population is in the age bracket 0 to 5, and there are as many Americans in the age bracket 40 to 45 as in the 0 to 5 age group. But rapid-growth countries still outnumber slow-growth countries, and so global population will continue to climb even as birth rates decline.

32

Now let us combine inevitable population growth with aspirations of people worldwide for a higher standard of living. One measure of quality of life, developed by the United Nations, is a Human Development Index (HDI). The HDI is based on life expectancy, educational level, and per capita gross domestic product. It is measured on a scale of 0 (poorest performance) to 1 (ideal performance). Figure 9 shows dramatically that energy consumption plays a determining role in the achievement of standard of living (HDI). This influence is particularly acute in the early stages of development in which the vast majority of the world's people finds itself. China and India (triangle symbols, Figure 9), the two most populous nations on Earth, which together comprise more than one third of the world's population, have per capita energy consumption of 10 and 40 gigajoules (1 GJ = 10^9 Joules), respectively. If these two countries increase their standard of living by following the curve in Figure 9, global consumption of energy

Figure 8. The population bomb. Human population traced from 1700 to the present. The Earth's population, about 1 billion at the time of the industrial revolution, currently is approaching 6.7 billion. Population distributions, in 5-year age brackets, are shown for rapid- and slow-growth countries (inset). Because rapid-growth countries exceed slow-growth countries, and because of population momentum, human population is predicted to rise to at least 9 billion before stabilizing.

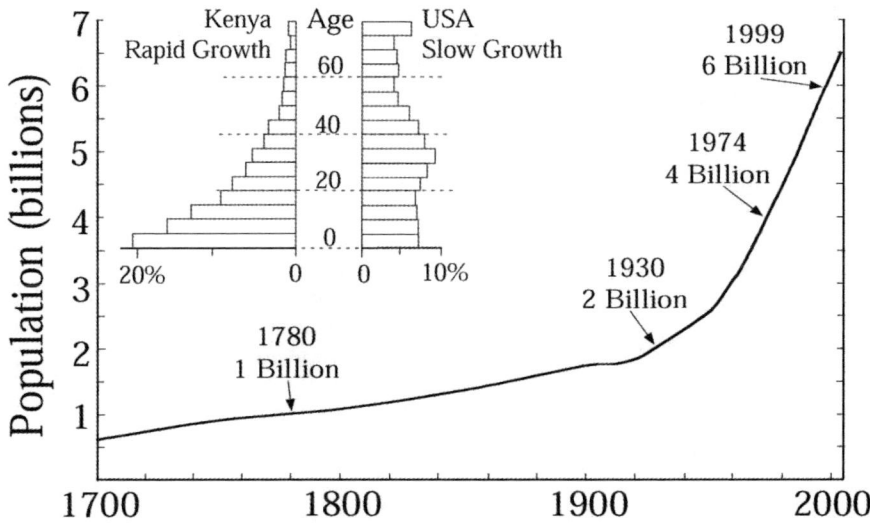

will increase dramatically. There are recent indications that such shifts already are taking place: China now consumes half of the world's cement, a third of its steel, and a quarter of its aluminum.

The influence of per capita energy consumption on the HDI begins to decline at about 50 to 100 GJ per capita. Thereafter, even with a trebling in energy consumption, the HDI does not increase significantly. A level of 100 GJ per capita could be sufficient to support a reasonable level of development if energy could be used efficiently. The three countries with the highest per capita energy consumption shown in Figure 9 are Canada, the United States, and Kuwait; these three countries have per capita energy appetites of 360, 350, and 320 GJ respectively.

How are these increasing demands for energy being met? Fossil fuels

Figure 9. Energy consumption drives standard of living. The human development index (HDI) for individual countries, a measure of living standard, shows a strong dependence on per capita energy consumption. Energy units are gigajoules ($1 \ GJ = 10^9 \ Joules$). Different symbols indicate country populations of 1 billion or more (triangles), 100 million to 1 billion (circles), and less than 100 million (diamonds). As high-population countries like Nigeria, India, Pakistan, China, and Indonesia increase their standards of living, the energy consumption path they choose will have major implications for global energy demands.

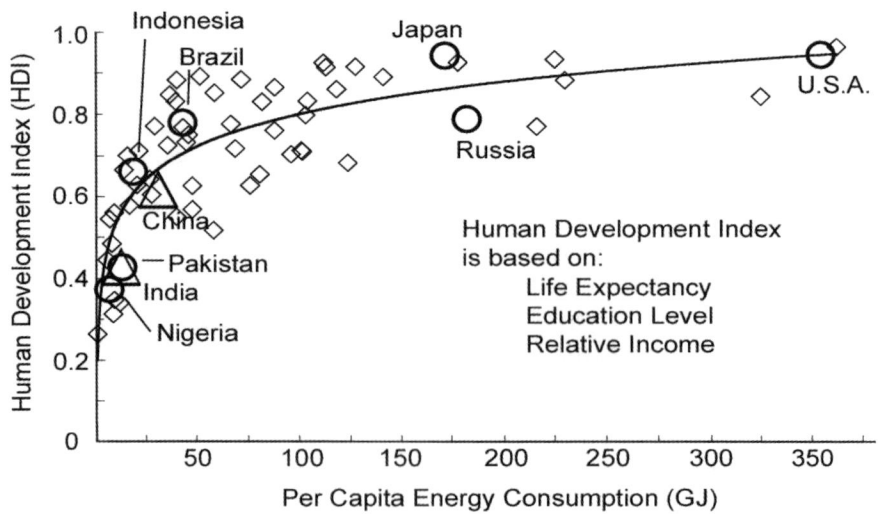

coal, oil, and natural gas account for 90 percent of energy consumption in the world today (Figure 10). That percentage has decreased only slightly since 1980, while the annual total energy consumption has increased 30 percent from 280 to more than 400 exajoules (1 exajoule = 10^{18} Joules of energy).

When taken together, the inevitable growth of human population, combined with individuals and societies having a legitimate desire to raise standard of living by consuming more energy, and the present reliance on fossil fuels for energy, the prognosis for more atmospheric carbon dioxide emissions and likely accelerated global warming is bleak.

But scenarios envisaged in the most recent IPCC report show that the

Figure 10. Global energy consumption. Global energy demand is growing at about 1.3 percent per year, slightly outstripping population growth. About 90 percent of human energy demand is satisfied by burning fossil fuels, releasing increasing amounts of carbon dioxide into the atmosphere and contributing to the enhanced greenhouse effect.

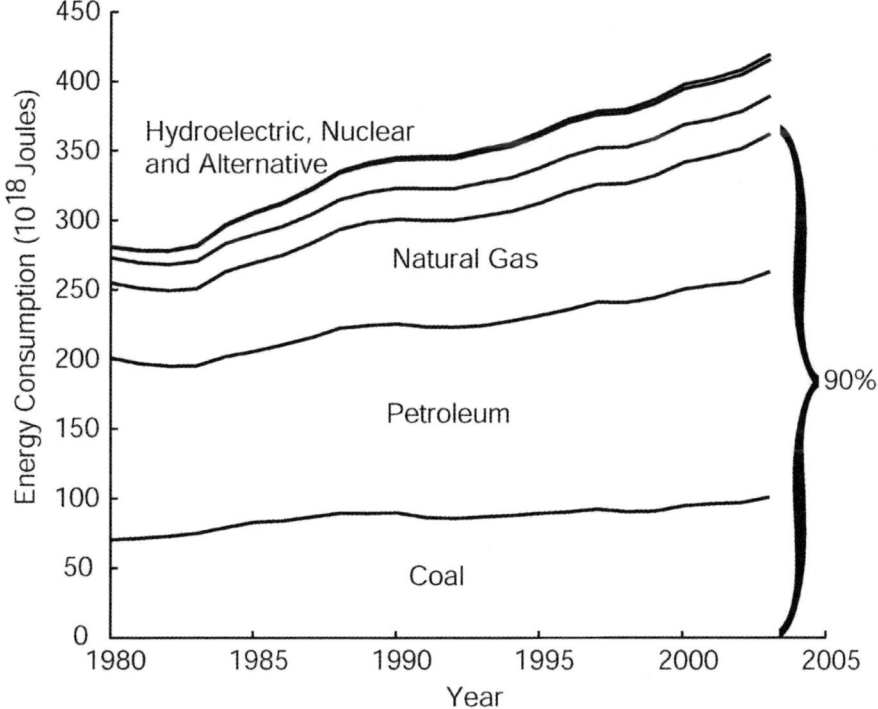

path we choose can make a difference in the longer term. Figure 11a shows various carbon emission scenarios between the year 2000 and 2100; equivalent atmospheric CO_2 concentrations for each of the emission scenarios are shown in Figure 11b. Drastic emission differences between the scenarios result from different assumptions about population growth, use of fossil fuel, technology, and a global sustainability ethic. For example, scenario A2 is "business as usual," with population growing to 15.1 billion by 2100 and continuing our heavy reliance on fossil fuel. Emissions rise from the current 8 billion tons (gigatons or Gt) of CO_2 annually to 30 Gt CO_2 in 2100. That scenario results in atmospheric CO_2 concentrations of 800 ppm in 2100, nearly three times the historic level prior to the industrial revolution. At the other extreme, scenario B1 follows a "balanced" path, with population rising to 8.7 billion by 2050 but decreasing to 7 billion by 2100. There is considerable emphasis on technology, alternative energy sources, and a global ethic to solve environmental problems. Emissions rise slightly until mid century but decrease steadily until 2100, when the emissions are 25 percent less than today. Even in this most optimistic scenario, atmospheric CO_2 increases slowly but steadily through the next century, reaching 560 ppm or a doubling of the historic CO_2 level (2 x 280 ppm) by 2100.

Figure 11. The atmosphere of the 21ˢᵗ Century. (a) Seven scenarios for carbon dioxide emissions combine projections for human population, technology, economics, and a sustainability ethic. (b) Each emission scenario results in a growth of carbon dioxide concentration for the next 100 years. (After IPCC.)

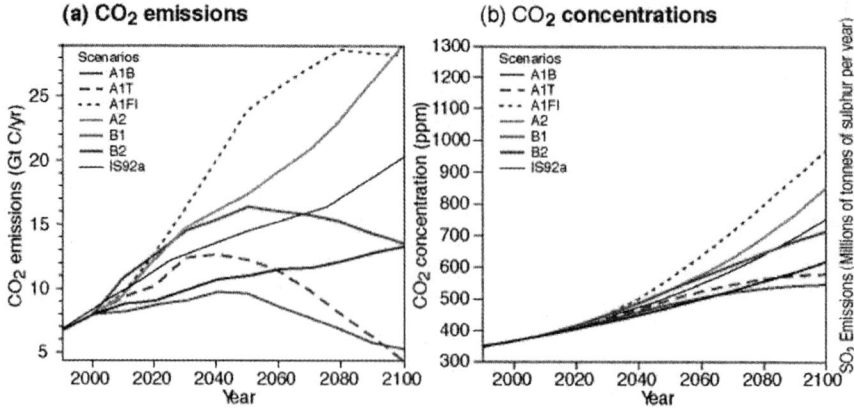

It is interesting to note that, even though the consequences of our current actions to curb or not to curb CO_2 emissions have extreme effects on CO_2 concentrations by the year 2100, the differences in concentrations do not emerge until about 2030. Do we have the foresight to take action now and not see immediate results of our actions for two decades?

What is the manifestation of greenhouse gas accumulation in regard to global warming over the next century? And how different is the future from the past? Figure 12 shows the global mean temperature projections to 2100 and, to place the change in context, reconstruction of surface temperature for the last millennium.

Relative to present-day temperatures, the Earth has experienced fluc-

Figure 12. Global temperatures—past, present, and future. Reconstructions of past temperatures are based on temperature proxies and borehole temperatures from seven different research teams. Each curve displays slightly different data sources. Uncertainty increases backward in time. However, all of the reconstructions provide a consistent picture of temperature variation over the last millennium. The instrumental record covers the period 1860 to present. Future projections of climate change show that over the course of the next century, surface temperature will continue to increase between 1° and 4° C, dependent on greenhouse gas emission scenarios.

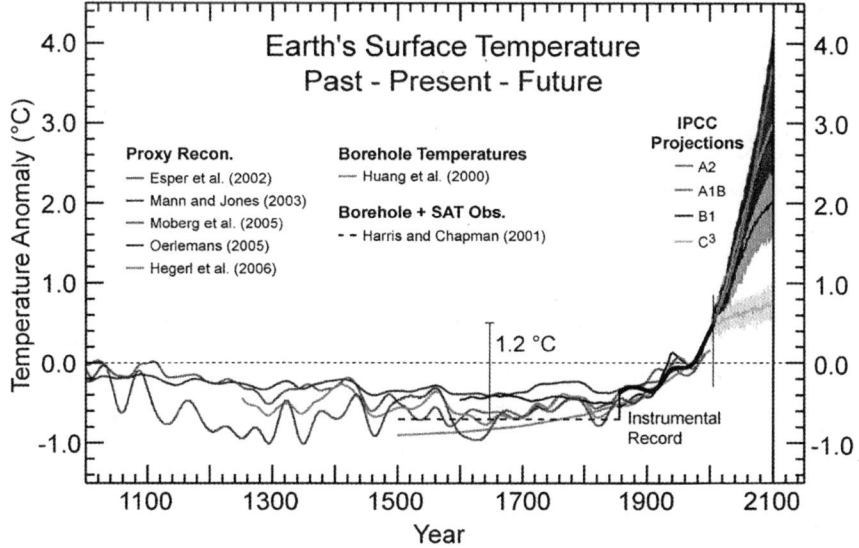

Figure modified from NRC (2006), IPCC (2007)

tuations with amplitude 0.5° C and decadal or longer periods for the last thousand years. This reconstruction is a significant update from earlier reconstructions, in particular Mike Mann's "hockey stick," which promoted much discussion about climate change but whose details have now been considerably modified. There is evidence (Figure 12) for a global medieval warm period around 1100 AD and a little ice age, particularly pronounced in Europe, from 1500 to 1700 AD, although not prominent in global reconstructions. The current warming episode of about 1° C in the last century, and up to 1.2° C since the industrial revolution is clearly emerging as an unusual temperature trend.

Figure 12 also shows predicted temperature changes to 2100 corresponding to the CO_2 emission scenarios in Figure 11. Global temperature increases, relative to the 1961-1990 mean temperature, for the A2 (business as usual), A1B (mixed) and B1 (balanced) story lines, are predicted to be 4°, 3°, and 2° C respectively. Measured from a baseline temperature at the start of current greenhouse gas accumulations, those temperatures are increased by another 0.6° C. The predictions are even more alarming when one realizes that regional warming, for example in the Arctic, may be three times as great as the mean.

The smooth increases in future temperatures shown in Figure 12, however, are unlikely to be realized in detail. There is no reason to believe that the chaotic components of climate that have caused the decadal variations in temperature over the last millennium should cease to exist in the future. Thus we should expect times of rapid temperature rise, such as the present, when greenhouse forcing is synchronized with natural variations, and times when greenhouse forcing is partially cancelled by natural variation for a limited period of time. These latter periods should not give rise to complacency because the greenhouse forcing is strong and long-lived (Figure 11).

Civic and Personal Choices

Let us return to the choices facing us. Some require global commitment, others require public leadership, and yet others rely on individual actions. The most general solution for the global warming dilemma requires changing the future projections of the curves shown in Figures 8, 9, and 10. First, population growth must be brought under control (Figure 8) so that the world's population stabilizes at 10 billion or fewer people, not more. Second, both developed and developing countries should aim to achieve a high standard of living in the sustainable per capita energy consumption range of 100 to 150 GJ (Figure 9). Canada and the United States should reduce their excessive energy consumption. Developing countries should try to maximize their HDI at the least growth in energy consumption, using models other than the United States and Canada for development. Third, creative talents of engineers and scientists need to be challenged into improved energy efficiency and fuel switching to reduce our dependency on fossil fuels (Figure 10).

Public leadership is important. To achieve the economic potential of improved energy efficiency and fuel switching, governments should provide a combination of targets and timetables, efficiency regulation and an array of market-based incentives that encourage businesses to make the necessary investments to reduce carbon dioxide emissions. Such measures could include: mandating high energy efficiency standards, retrofitting buildings to conserve energy, reducing subsidies that distort energy prices, developing carbon taxes or market-based measures such as tradable emissions permits (discussed elsewhere in this volume), encouraging fuel switching to less carbon intensive fuels, developing renewable energy sources, working with automakers to encourage use of more energy efficient vehicles, and assisting municipalities with planning that minimizes vehicle use. All these measures will have greatest effect if implemented in a timely manner. Some studies indicate that policies to reduce greenhouse gas emissions produce economic benefits greater than their costs. Policies encouraging energy efficient processes and renewable energy technologies are a bridge to the knowledge-based economies of the 21st century.

Individuals also can make a difference. Strategies include: using a fuel-efficient car and driving less; living closer to work and walking or riding a bicycle; buying local and seasonal goods to reduce consumption of energy for transportation; making sure that one's house is well-sealed and insulated to reduce heating in the winter and cooling in the summer; using compact fluorescent light bulbs and energy-efficient appliances; planting trees and shrubs; composting and recycling; and, through the democratic process, encouraging elected officials to deliver policies that properly take the environment into account. After all, if people can cause global warming, people can stop it.

The global warming debate, finally, may force us to develop a more global and less insular perspective of planet Earth's future. Climate change problems and issues are complex. The scales are global and the time scales are decades to centuries. The challenge, and overall goal, is appropriate environmental stewardship of the planet. Global warming may just be the alarm that brings us, albeit with much debate, to action.

SUMMARY VIEW OF CLIMATE CHANGE

William R. Cotton

IT has become increasingly unpopular to criticize the notion that global climate change is due to human-produced increases in carbon dioxide or CO_2 concentrations. The fact that Al Gore and the Intergovernmental Panel on Climate Change (IPCC) shared the 2007 Nobel Prize for Peace is an example of how politically charged this concept is. In spite of the fact that I am quite green in my lifestyle (I own a Toyota Prius, ride a bicycle to and from work, paddle or sail boats, and fly a sailplane), as a scientist I feel compelled to critically examine the factors that contribute to global climate change, including CO_2 contributions as well as other factors [see a review in Cotton and Pielke (2007)].

The Evidence for Global Warming

I think there is considerable evidence that the planet earth is warming. Furthermore, the concentrations of CO_2 are also increasing at alarming rates. The question is, are these cause-and-effect, or is the planet warming for other reasons? The answer is not trivial. I have been associated with weather modification studies for over 40 years. There are several hypotheses for human-induced changes in weather, especially precipitation. Many of these hypotheses make physical sense and are treated quantitatively in models. But after over 50 years of research, it has been difficult to establish that human-induced changes have caused observed changes in precipitation on the ground. The reasons for this failure are many but are most likely tied to the fact that the forcing we induce by cloud seeding is small compared to the forcing produced by nature. Thus, it is difficult to separate human-caused changes from changes due to many other factors that we can lump into the "natural variability" of the system.

The same can be said about global climate variability. Is the observed warming trend due to "natural variability" or a direct consequence of human-produced enhanced concentrations of CO_2?

The so-called "hockey stick" paper of Mann et al. (1998) provides the strongest evidence that the current period of global warming is unprecedented over the last 1,000 years or so. But this paper has been criticized by a number of authors [McIntyre and McKitrick (2003), for example] as having major problems in the statistical treatment of the data. Moreover, the hockey stick results are inconsistent with surface temperatures inferred from reconstructions of surface temperatures for the Sargasso Sea [Robinson et al., (2007)]. The hockey stick data suggest that the warming period we are experiencing has been going on for over 300 years, since the end of the Little Ice Age, and that the Medieval Climate Optimum period 1,000 years ago was much warmer. There is circumstantial evidence that the climate in Greenland, for example, was much warmer than now during the Medieval Climate Optimum period because the glaciers were much reduced in coverage and the seas were more open to navigation.

The Greenhouse Gas Concept

Carbon dioxide is clearly a major absorber of long-wave radiation and therefore contributes to so-called greenhouse gases. But we need to keep in mind that CO_2 is not the major greenhouse gas, because water vapor has that distinction. Thus, much of the greenhouse warming in models is due to feedbacks that involve a spin-up of the hydrologic cycle, leading to higher concentrations of water vapor in the atmosphere, which then contributes to most of the greenhouse warming. The spin-up of the hydrologic cycle remains a major source of uncertainty in global models, because it depends on the response of the ocean system and of land-surface systems (that is, responses in soils and the biosphere). Moreover, clouds are very important absorbers of long-wave radiation and determine the albedo of planet Earth. Low clouds tend to enhance the earth's albedo (a cooling effect), while high clouds tend to absorb more long-wave radiation and therefore serve as greenhouse warmers. Because models depend on rather crude parameters for clouds, it is still uncertain how clouds respond to a warming planet and to a spin-up of the hydrologic cycle. The question is, are there more high clouds versus low clouds in a warming planet, and how does the cloud distribution vary with latitude? Increased cloud cover at high latitudes contributes to a warming trend in the Arctic because the surface energy budget at high latitudes is dominated by long-wave radiation, even in the summer months.

In fact, another important feedback in global models is the ice-albedo feedback, in which the models respond to CO_2-induced warming by reducing snow-ice-glacier coverage, which leads to a reduction of planetary albedo and therefore a positive warming feedback. But much of the reduction in snowpack and glacier coverage is not due to changes in albedo. As mentioned above, the high-latitude energy budget is dominated by long-wave radiation. Thus, melting of snowpack and glaciers is mainly a result of changes in cloud cover (increasing cloud cover contributes to warming) and to advection of heat into higher latitudes by atmospheric and ocean circulations.

Factors Contributing to Global Climate Change

While greenhouse gases, especially water vapor, are a major contributor to the habitability of planet Earth, is the variability of these gases the dominant contributor to climate change? That is the one trillion dollar question.

What are some of the other competing processes that change the forcing of our climate system? These are reviewed in Cotton and Pielke (2007) and include the following:

- Changes in solar luminosity and orbital parameters;
- Changes in surface properties; and
- Natural and human-induced changes in aerosols and dust.

Changes in earth-orbital parameters, the so-called Milankovitch cycle [Imbrie and Imbrie, (1979); Berger (1982)] are believed to be responsible for the onset of ice ages. But, those changes cannot explain the current warming trend because they seem to predict that we will be moving into an ice age in the next 5,000 years. While there is evidence of a small variation in the sun's irradiance, the amount of variability is too small to account for recent climate variations or even those over the last 1,000 years. While there have been many studies suggesting statistical correlations between varying solar parameters and Earth's climate, the physical causes of those correlations are for the most part not well founded.

Variations in land surface properties affect the planetary albedo and alter the surface energy budget enough to change the Bowen ratio. Human activity contributes to changes in surface properties through agricultural land use and urbanization. Moreover, changes in land use and vegetation respond to climate changes in a nonlinear way, thus altering both the planetary albedo and the surface energy budget. While changes in land surface properties are a significant contributor to the planetary energy budget, they probably do not rank as high as greenhouse warming [IPCC (2007)]. Nonetheless, I think that the contribution from changes in land use are larger than the IPCC's estimates.

Cotton and Pielke (2007) devoted an entire chapter to human-induced changes in aerosols. That chapter considered both the direct and semi-direct effects of aerosols and dust, as well as indirect effects, that alter the earth's albedo and hydrologic budget through alterations in cloud properties. Large uncertainties exist in estimating the consequences of aerosols on climate largely because of the fact that a major contributor is related to cloud processes that are poorly represented in global climate models. Still, it is generally believed that human-induced changes in aerosols contribute to a net cooling in the climate system, which offsets warming effects by roughly one-third the amount of greenhouse gas warming [IPCC (2007)]. This process sometimes is referred to as "global dimming."

Since writing my portion of Cotton and Pielke (2007), I have come to believe that the major wild-card variable in the climate system is naturally produced aerosols and specifically aerosols induced by volcanic activity. When we wrote that book, I thought that volcanic activity was purely random. Since then, however, I have come across a series of papers by Reid Bryson and colleagues [Bryson and Goodman (1980a, 1980b); Bryson (1982, 1989); Goodman (1984)]. I am now convinced that volcanic activity is modulated by sun-moon-earth tidal variations, thus introducing a degree of predictability into the climate system. Under this scenario, periods of global warming can be attributed to periods of very low volcanic activity, like between 1920 and 1940 [Robock (1979)] and the Medieval Climate Optimum period. On the other hand, periods of extensive cooling like the Little Ice Age were periods of maximum alignment of the sun-moon-earth tidal forcing, which contributed to very active episodes of volcanic activity and global cooling. The consequence of this hypothesis would be that forecasts of global greenhouse gas warming would be at the mercy of climate variability due to volcanic activity. Periods of greater than normal volcanic activity could completely override or mask the forcing by greenhouse gases.

Thus I still ask the one trillion dollar question: Is the current warming trend due to greenhouse gas warming, just to a period of below-average volcanic activity, or to some other source of natural climate variability?

Climate Engineering

More and more we encounter suggestions for altering the warming tendency attributed to greenhouse gases by climate engineering. In my opinion, engineering the climate system is very dangerous, because we cannot predict the long-term climate responses to our interventions, no more than we can predict the consequences of greenhouse gas emissions. In fact, we could induce a global cooling response and find ourselves in the midst of a period of global cooling due to increased volcanic activity or some other source of cycle-reinforcing natural variability, and drive the earth into an ice age whose consequences could be far worse than global warming!

Summary

There are strong indications that our global climate is warming. But the question is, is the warming due to anthropogenic greenhouse gases, or is it due to some other forcing mechanisms (or their transient absence) and natural variability. As human population on Earth continues to increase, the chances of human-induced changes in climate due to greenhouse gases, aerosol pollution, or alterations in land use become increasingly likely. Thus, rather than consider climate engineering, we should devise methods of encouraging the reduction of population growth through economic and quality-of-life incentives.

References

Berger, W.H., 1982: "Climate steps in ocean history—lessons from the Pleistocene," in *Climate in Earth History*, W.H. Berger and J.C. Crowell, panel co-chairmen. Washington, DC: National Academy Press, pp. 43-54.

Bryson, R.A., 1982: "Volcans et climat," *La Recherche*, 13 (135), 844-853.

Bryson, R.A., 1989: "Late quaternary volcanic modulation of Milankovitch climate forcing," *Theor. Appl. Climatol.*, 39, 115-125.

Bryson, R.A., and BM. Goodman, 1980a: "Volcanic activity and climatic changes," *Science*, 27, 1041-1044.

Bryson, R.A., and BM. Goodman, 1980b: "The climatic effect of explosive volcanic activity: Analysis of the historical data," NASA Conference Publication no. 2240. *Proceedings, Symposium on Atmospheric Effects and Potential Climatic Impact of the 1980 Eruptions of Mount St. Helens, November 18-19, 1980*, Washington, DC.

Cotton, W.R., and R.A. Pielke, 2007: *Human Impacts on Weather and Climate,* 2nd edition. Cambridge University Press.

Goodman, B.M., 1984: "The climatic impact of volcanic activity." Ph.D.

dissertation, Department of Meteorology, University of Wisconsin-Madison, 245 pp.

Imbrie, J., and K.P. Imbrie, 1979: *Ice Ages: Solving the Mystery*. Short Hills, NJ: Enslow Publishers, 224 pp.

McIntyre, S. and R. McKitrick, 2003: "Corrections to the Mann et al. (1998) proxy data base and the Northern Hemispheric average temperature series," *Energy & Environ.*, 14, 751-771.

Mann, M.E., R.S. Bradley, and M.K. Hughes, 1998: "Global-scale temperature patterns and climate forcing over the past six centuries," *Nature*, 392, 779–787.

Intergovernmental Panel on Climate Change (IPCC), 2007: *Climate Change 2007—the physical basis: Working Group I contribution to the fourth assessment report of IPCC*. [S.D. Solomon et al., eds.] New York, NY: Cambridge University Press, 996 pp.

Robock, A., 1979: "The 'Little Ice Age': Northern Hemispheric average observations and model calculations," *Science*, 206, 1402-1404.

Robinson, A.B., N.E. Robinson, and W. Soon, 2007: "Environmental effects of increased atmospheric carbon dioxide," *J. Amer. Physicians Surgeons*, 12, 79-90.

COMMENTS ON HURRICANES AND GLOBAL WARMING*

By William M. Gray

THERE is another side to the scientific climate change dispute of which most of you probably are unaware. The media naturally have come to me for comments on how I interpreted the very active and damaging hurricane seasons of 2004-2005. That is because I have been issuing Atlantic seasonal hurricane forecasts for 23 years and studying hurricanes for nearly 50 years. I have been given wide media coverage for my seasonal hurricane forecasts, and I feel an obligation to try to set the record straight on how to interpret these two recent years with very active hurricane seasons.

The statements by Emanuel, Trenberth, Holland, Webster, and Curry—whom I henceforth will refer to as the "Gang of Five"—have caused needless alarm and worry among hundreds of thousands of coastal residents in the United States and abroad. They have not recognized that they do not have a license to yell "Fire!" along a crowded eastern United States coastline.[1]

Unprecedented increases in U.S. hurricane damage were anticipated by me and other hurricane specialists for a number of years prior to 2004-2005 when we entered a new phase of the Atlantic multi-decadal oscillation (AMO) in 1995. This expectation was due to the large coastal population increases since the last active Atlantic hurricane period of the 1940s and 1950s. Most of us who have studied and forecasted hurricanes for a long time believe that the U.S. hurricane damage of 2004-2005 was well within the range of natural variability.

But suddenly in 2005, out stepped the Gang of Five to proclaim confidently, with no valid evidence, that human-induced global warming played a significant role in the 2004-2005 U.S. hurricane destruction. What did they know that the rest of us who have more background in hurricane climate and forecasting did not know? How could they be so confident of their results that they felt at ease in alarming the public the way they did? Of the Gang of Five, only Emanuel has any prior research experience in the

---

* This was prepared from a November 13, 2006 paper by Professor Gray on the same subject as his conference topic.

climate-hurricane association.

The Gang of Five appears not to have comprehended the implications of their papers within a much larger and more serious arena than the typical tropical cyclone (TC) papers we publish within our peer group. Most of the TC papers do not go much beyond our own group and can do little harm. As well-established senior scientists, they (the Gang of Five) had a professional responsibility—beyond their own interests—not to falsely alarm the public at such an unusual period of massive U.S. hurricane destruction. I would not have made comments on the *Nature* and *Science* papers that they published if the scientists had not become so public and accepted as so believable by a public already brainwashed about global warming. In the usual case there would have been little problem. Our own peer group gradually would have sorted out the faulty data, and results would not have been discussed so extensively in the media.

Senior meteorologists (such as the Gang of Five) should have developed, over the years, a better intuitive feeling about how the atmosphere-ocean interaction functions. They should have been more surprised and far more critical of their data which, if correct, had profound meaning for global climate change and future hurricane destruction. As well-known and respected scientists, they should have been more cautious and should not have surrendered their objectivity and their responsibility as easily as they have done.

The Gang of Five encouraged the media to trumpet their findings with no concern or sense of responsibility for the effect their comments might have on many millions of coastal residents of the United States and abroad who cannot judge the validity of their comments. Their continuing comments have affected many of the decisions of the insurance industry, investment groups, business owners, etc., besides causing unneeded alarm and worry among millions of hurricane-vulnerable homeowners.[2]

When data to the contrary has been presented to the Gang of Five, they do not back away but instead criticize the person presenting the alternate view. The alleged personal attacks that the Gang of Five say that I have made on them by trying to tell the truth about their data errors are (from my point of view) much less egregious than the media blog reports of what outlandish things they have been saying about me in these venues. Given their responses to me, I can understand why younger and more vulnerable colleagues would have been reluctant to confront them on this topic.

It appears that no reasonable amount of data or valid counterarguments will dissuade them. For example, in Holland's talk to a congressional committee on October 27, 2006, he said,

> The strong relationship between increases in storm and hurricane numbers and increases in SSTs [sea surface temperatures] leads to the inescapable conclusion that the majority of current hurricane activity is a direct result of greenhouse warming.

Emanuel continues to muddy the waters with outrageous claims that the multi-decadal oscillation (AMO) does not exist. Trenberth gives information on the percent of Atlantic hurricane activity in 2005 that is due to global warming and implies that humans are a significant part of the observed hurricane activity of that year. He does not tell how he knows this. None of the Gang of Five appears to know what they are talking about. All of their statements can and will in time be refuted.

It is important to realize that the Gang of Five's statements do not fit into the normal esoteric kind of data crunching or modeling that typically remains within our peer group. They are having an undeserved impact on the public's perception of hurricanes and future hurricane destruction.

When the media repeatedly asks me why the Gang of Five would so exaggerate human influences on hurricane destruction when the evidence does not support this assessment, what am I to say? Their retort to my criticism is not to present more and better evidence to better back their claims, but instead to attack me with the argument that I am engaging in character assassination.

Rather than complaining about me to my department head, Rick Anthes should be taking some responsibility for not reining in Trenberth's and Holland's wild and wrong assertions about humans causing more intense hurricanes. As a public institution, UCAR [University Corporation for Atmospheric Research] ought to feel some obligation not to falsely alarm the public. Anthes has a strong background in tropical cyclones, and he is well qualified to speak out. He and I sat next to each other at dinner in Miami, Florida, in May 2006. I told him of UCAR's problems with Trenberth and Holland, but he did not wish to talk about it. I judge him, unfortunately, to be a silent supporter of the Gang of Five's claims.

The Gang of Five's assertions have become the laughing stock of most of the experienced hurricane researchers whom I know and respect. Their ideas would totally collapse before a group of us who really know something about hurricanes. It is ludicrous to hear Judy Curry and Peter Webster, who have no previous hurricane experience, discussing the coming CO_2-induced hurricane increases on national television and radio.

As highly talented and credentialed researchers, the Gang of Five apparently believe that their reputations will be able to persuade the uninformed of their assertions and that they will be able to sell Washington funding sources on their views—they apparently believe that they can just run over their critics.

Here are a few of the Gang of Five's misrepresentations:

Trenberth—"At the same time, there has been a surge over the past 35 years in the number and proportion of intense hurricanes." (Not true.) It's highly likely that greenhouse gases are partly to blame and that the trend will continue. (Likely not true.)

"The prospects are for more intense storms, heavier rainfall and flooding, and more coastal damage. Global warming also boosted the amount of rainfall in several of the most powerful 2005 hurricanes—including Katrina and Rita—by 7 percent." (How can he possibly estimate a number like 7 percent?)

Holland—He compares increases in named storms from 1900 to the present, including before aircraft/satellites were used. Many of the earlier period cyclone numbers and those in the central Atlantic consequently were missed. His upward trend in named storms means nothing. Holland knows better than this but persists in this charade.

Emanuel—He uses the cube of the tropical cyclone's [TC's] maximum winds over data deficient oceans. Cubing poor data leads to wild errors. He has misrepresented past data sets to get desired TC upward curve results. He does not understand the Atlantic multi-decadal data set and says that it does not exist.

Webster and Curry—They show Category 4-5 cyclones increasing from the 1970s. I visited all global tropical cyclone centers in 1978-1979 as part of a World Meteorological Organization survey trip, and I have stated many times that there is no way that these centers could have distinguished Category 4-5 cyclones from Category 1-2-3 cyclones. The satellites for these centers often did not work, and forecasters were not trained to use them as they currently are. These centers naturally underestimated the number of Category 4-5 cyclones in these earlier years. An upward trend in Category

4-5 cyclones from the 1970s means nothing.

My motivation in all this explanation is only to try to help maintain the integrity of American science which, in my view, has been badly compromised by the global warming issue and now recently by the issue of global warming causing more frequent and more intense hurricanes. Having received federal support for my hurricane research and forecasting endeavors for nearly 50 years, I feel I have an obligation to speak out on issues involving my expertise, particularly when statements are made which are contrary to everything I have learned over my long career. I would feel guilty if I said nothing.

Attached is an Appendix of my alternate view of reality.

Appendix: Another View of Reality

Although global surface temperatures have increased over the last century and over the last 30 years, there are no reliable data available to indicate increased hurricane frequency or intensity in any of the globe's seven tropical cyclone basins. Meteorologists who study tropical cyclones have no valid physical theory as to why hurricane frequency or intensity would necessarily be altered by small amounts ($< \pm 0.5°C$) of global mean temperature change.

In a global warming or global cooling world, the atmosphere's upper air temperatures will warm or cool in unison with the sea surface temperatures. Vertical lapse-rates will not be significantly altered. We have no plausible physical reasons for believing that Atlantic hurricane frequency or intensity will change significantly if global ocean temperatures continue to rise. For instance, in the quarter-century period from 1945 to 1969, when the globe was undergoing a weak cooling trend, the Atlantic basin experienced 80 major (Category 3-4-5) hurricanes and 201 major hurricane days. By contrast, in a similar 25-year period of 1970-1994, when the globe was undergoing a general warming trend, there were only 38 major hurricanes (48 percent as many) and 63 major hurricane days (31 percent as many). Atlantic sea-surface temperatures and hurricane activity do not necessarily follow global mean temperature trends.

The most reliable long-period hurricane records we have are the measurements of U.S. landfalling tropical cyclones since 1900 (see Table 1). Although global mean ocean and Atlantic surface temperatures have increased by about 0.4° C between these two 50-year periods (1900-1949 compared with 1956-2005), the frequency of U.S. landfall numbers actually shows a slight downward trend for the later period. If we chose to make a similar comparison between U.S. landfall from the earlier 30-year period of 1900-1929 when global mean surface temperatures were estimated to be about 0.5° C colder than they have been the last 30 years (1976-2005), we find exactly the same U.S. hurricane landfall numbers (54 to 54) and major hurricane landfall numbers (21 to 21).

We should not read too much into the last two hurricane seasons of 2004-2005. The activity of these two years was unusual but well within natural bounds of hurricane variation. Between 1966 and 2003, U.S. major hurricane landfall numbers were below the long-term average. Of the 79 major hurricanes that formed in the Atlantic basin from 1966 to 2003 only 19 (24 percent) of them made U.S. landfall. During the two seasons of 2004-2005, seven of 13 (54 percent) came ashore. This is how nature sometimes works. What made the 2004-2005 seasons so unusually destructive was not the high frequency of major hurricanes but the high percentage of major hurricanes that were steered over the U.S. coastline. The unanticipated breeching of the New Orleans levees likely doubled or tripled the damage caused by Katrina. The major U.S. hurricane landfall events of 2004-2005 were primarily a result of the favorable upper-air steering currents that were present during those two hurricane seasons.

Although 2005 had a record number of tropical cyclones (27 named storms, 15 hurricanes, and 7 major hurricanes), this should not be taken as an indication of something beyond natural processes. There have been several other years with hurricane activity comparable to 2005. For instance, 1933 had 21 named storms in a year when there were no satellite or aircraft data. Records of 1933 show that all 21 named storms had tracks west of 60° W longitude, where surface observations were more plentiful. If we eliminate all the named storms of 2005 whose tracks were entirely east of 60° W and therefore may have been missed, given the technology available in 1933, we reduce the 2005 named storms by seven (to 20)—about the

Table 1. U.S. landfalling tropical cyclones by intensity during two 50-year periods.

Years	Named Storms	Hurricanes	Intense Hurricanes (Cat. 3-4-5)	Global Temperature Increase
1900-1949 (50 years)	189	101	39	+0.4° C
1956-2005 (50 years)	165	83	34	

same number that occurred in 1933.

Using the National Hurricane Center's best track database of hurricane records back to 1875, six previous seasons had more hurricane days than the 2005 season. These years were 1878, 1893, 1926, 1933, 1950, and 1995. Also, five prior seasons (1893, 1926, 1950, 1961, and 2004) had more major hurricane days. Finally, five previous seasons (1893, 1926, 1950, 1961, and 2004) had greater Hurricane Destruction Potential (HDP) values than 2005. HDP is the sum of the squares of all hurricane-force maximum winds and provides a cumulative measure of the net wind force generated by a season's hurricanes. Although the 2005 hurricane season was certainly one of the most active on record, it is not as much of an outlier as many have indicated.

Most of my tropical cyclone colleagues who have spent years forecasting and studying hurricanes do not subscribe to the alarmist views of those saying we should expect hurricanes to become worse due to human-induced global warming. We believe that the Atlantic basin is currently in an active hurricane cycle. This active cycle is expected to continue for another decade or two, at which time we should enter a quieter Atlantic major hurricane period like the one we experienced during the quarter-century periods of 1970-1994 and 1901-1925. Atlantic hurricanes go through multi-decadal cycles. These cycles have been observationally traced back to the mid-19th century and inferred from Greenland paleo-ice core temperature measurements that go back thousands of years.

Endnotes

[1] Editor's note: This controversy arose at the 2006 annual meeting of the American Meteorological Society in Atlanta, Georgia. It is well documented in Valerie Bauerlein, "Cold Front: Hurricane Debate Shatters Civility of Weather Science," *The Wall Street Journal*, February 2, 2006, p. A-1. The five scientists to whom Prof. Gray refers, in the order named in that article, are Greg Holland, National Center for Atmospheric Research; Judith Curry, Georgia Institute of Technology; Kevin Trenberth, National Center for Atmospheric Research; Kerry Emanuel, Massachusetts Institute of Technology; and Peter Webster, Georgia Institute of Technology.

[2] Here is an excerpt from a letter I received recently from a large Florida hotel operator: "As the owner-operator of seven hotels in Central Florida

with approximately 6,500 rooms, I must confess that in recent months I have become quite concerned and somewhat curious as to why we have witnessed such a severe decline in occupancies for the months of August-October. Indeed, when compared to the past 30 years that we have been in business here in Central Florida, the three months in question have declined to the lowest levels of occupancy we have ever experienced. Concerned about this rather strange phenomenon, we decided to conduct an extensive survey of former guests in an attempt to discover the reason for the decline. We surveyed hundreds of guests who previously had stayed with us during these very same months in past years, and we asked them why they chose not to return this year. The results of our survey were stunning. Approximately 75 percent of those surveyed indicated that they chose not to return because of their concern about hurricanes. No doubt, hundreds of thousands of would-be visitors to Florida decided not to come presumably because of your rather dire forecast. In Central Florida alone we have lost tens of millions of dollars in revenue, as well as hundreds, perhaps thousands of jobs. If one were to extrapolate from our experience in Central Florida to the state as a whole, possibly billions of dollars have been lost, and tens of thousands of hospitality workers have either lost their jobs or had their hours dramatically reduced."

PANEL 3—THE BEST OF TIMES, THE WORST OF TIMES

James I. Mills

INITIALLY, I believed that in my capacity as Discussant for Panel Three, "Scientific Analysis of Global Climate Change," I would hear contrasting interpretations of the science of global climate change. I was fairly certain that by panel's end I would have witnessed a reasonably balanced presentation of scientifically framed worldviews, each supporting one of two hypotheses: either we, and the planet upon which we rely for sustenance, are poised for the best of times, or else the worst.

In fact, my expectations were only partly fulfilled. Gordon Michaels, representing Oak Ridge National Laboratory, argued neither for nor against a scientific argument for climate change. Instead, he presented a thorough, neutral, and sobering synopsis of the state of energy technologies that developed and developing societies have or may have access to as they struggle to: 1) curb greenhouse gas emissions, assuming they are committed to doing such a thing; and 2) provide the power required by an exponentially increasing global industrial output. At the end of this presentation, my economic and thermodynamic intuition was rewarded: there is no free lunch. All technologies or combinations thereof with promise to provide the silver rocket upon which we ride to a near-term future of "green" energy economies are currently too expensive, too inherently risky, too constrained in infrastructure, or at too early a stage of development. Michaels, without directly addressing the science of climate change, essentially suggests that technologies do in fact exist or have been imagined that could, some day, provide the underpinnings of a sustainable world. Whether or not they will arrive in time, however, is highly dependent upon the uncertain rapidity with which global economies continue to expand, how quickly markets can respond, and to what degree governments are willing or able to commit. Uncertainties abound, and growth carries on.

In contrast to George Michaels, Professor William Gray assured us that we have nothing to fear but fear itself: global warming, he asserts, is not a problem. Professor Gray is well-known for his skepticism about human-induced global warming, and his presentation focused upon his hypothesis that a natural cycle of ocean water temperatures—related to the amount of salt in ocean water—is responsible for the global warming that

55

he acknowledges has taken place. Professor Gray additionally asserts that this period of natural warming will be followed by a period of cooling, and thus, at least over the long term, we should neither concern ourselves with nor respond to current warming trends.

Although it could be argued that Professor Gray defends an increasingly contrarian position, his skepticism is an absolutely necessary contribution to the development of a comprehensive understanding of global climate change. Robert Merton, as one of many possible examples, makes the case clearly when he writes, "Most institutions demand unqualified faith; but the institution of science makes skepticism a virtue."[1]

While most of us can agree that Professor Gray makes a contribution with the challenge of his skepticism, all of us, I trust, will question the value of his undocumented claim that current theories of human-induced global warming are supported predominantly by scientists afraid of losing grant funding. Undocumented, this claim stands as destructive and fragmenting gossip, and accomplishes no more than the same.

In the light of the two presentations offered in Panel Three, I conclude predictably that climate change is all things to all people. To some the real or imagined looming of global warming represents an opportunity to search for and implement new technologies that will allow human enterprise to carry on sustainably. To others, climate change is nothing but a fabrication by those who, for largely undeclared and unfathomable reasons, desire to slow human progress and disrupt the enterprise of human progress.

Undoubtedly one of the reasons for the range of opinions and the dramatic emotions that fuel the controversy is the fact that the problem of global climate change is a problem embedded in a complex system controlled by innumerable feedback loops, some native, ancient, and subtle, some newer, anthropogenic, and potentially more profound, and all of which make long term projections—and, indeed, short term measurement—fraught with uncertainty. The complexity of this system, in fact, casts the whole problem into a class that we refer to as "wicked."

The very nature of wicked problems, in fact, goes a long way toward offering an explanation of why, even within the constrained borders of Panel Three, the global climate change debate is sure to offer such a broad range of disconnected foci—in our own panel, for example, the jump from discussions of emergent technologies to claims of fraudulent science offers, at the very least, a dramatic shift of theme if not an outright discontinuity.

Rittel and Webber's (1973) formulation of wicked problems[2] specifies a number of characteristics, the most critical of which I paraphrase as follows:

1. There is no definitive formulation of a wicked problem.
2. Solutions to wicked problems are not true-or-false, but better, or, worse.
3. Because the variables that control both the dynamics of wicked problems and our attempts to solve or "tame" them are so many, complex, and interdependent, and because great lag times can often exist between the imposition of a solution and a complex system's response to that imposition, there are no convenient tests of a solution to a wicked problem.
4. Every wicked problem, again because of the complex and dynamic nature of "wickedness," is unique, and thus previous experience from problems that we might consider similar is of little value.
5. Every wicked problem can be considered to be a symptom of another problem—thus, global warming may in fact be symptomatic of an even larger problem that, at its core, is more about the dynamics of historic trends in human population and economic growth then about carbon emissions.
6. The existence of inevitable controversy surrounding our interpretation of and approach to a wicked problem can be explained by the fact that every stakeholder will, understandably, view the problem from a unique perspective, and will tend to form hypotheses appropriate to those perspectives.

Jeff Conklin offers a similar but even more succinct set of wicked problem characteristics.[3] According to Conklin, the four characteristics of wicked problems, again paraphrased, are as follows:

1. The problem is not understood until after formulation of a solution— thus, our attempts to curb carbon emissions might force us, ultimately, to focus more on the problem of our inefficient transportation systems than upon a problem of global climate change, per se.
2. Stakeholders have radically different worldviews, or mental models, and different frames for understanding the problem—one person's fear that a quick response to the incompletely understood phenomenon of global warming might threaten industrial output is another person's

fear that a lack of such a response might threaten the systems that sustain the earth.

3. Constraints and resources available to solve the problem change over time, and therefore today's best solution is not necessarily tomorrow's.

4. As a result of the first three characteristics, wicked problems are never *solved*, but only *managed*.

Conklin's characterizations, in particular, give us insight into the nature of the current global climate change debate. If, for example, our proposed solution is curtailment of carbon emissions, then the problem that we actually pose is one of carbon emissions, and not, specifically, climate change, although it is the nature of wickedness that the solution to one problem may well impact another. If, on the other hand, we suspect that the nature of the problem is far more complex than that suggested by a focus on the reduction of carbon emissions alone—as complex as that proposed solution actually is—then we open ourselves to an awareness that complex problems warrant complex solutions. Then we become more likely to embrace a systemic solution approach that excludes none of the critical system components, including the nature of global energy economies, the nature of global production and consumption patterns, the nature of human population dynamics, and the degree to which we understand the potential fragility of the earth's climate feedback control mechanisms. This last point may be especially important. While we might argue, as Professor Gray, for example, apparently does, that native climate feedback systems have evolved in ways that lead to an inherent stability over time, the degree to which relatively rapid, anthropogenic perturbations to those systems might distort or alter feedback is, I would suggest, not comprehensively understood.

If we accept the fact that global change—and, indeed, global environmental change in general—is a wicked problem, then the emergent literature of wicked problems both addresses ways to manage these problems and cautions us about attempting to tame such problems. Examples of approaches to taming—well-meaning, perhaps, but futile all the same—include:

1. Lock down the problem definition—declare, for example, the problem to be fossil fuel emissions and focus on single solutions such as nuclear power or emission caps.

2. Assert that there is no problem—claim that global climate change is natural, has nothing to do with human economic activity or population dynamics, and that, therefore, a human response is neither required nor prudent.

If the above approaches sound all too familiar, and if their continued defense represents a source of concern, then it might be prudent for us to consider more appropriate ways of responding to global environmental and climate change.

As we search for a path that will lead us to an appropriate response to wickedness, we should keep well in mind the idea that wickedness is a function of both complexity and social fragmentation. Complex problems are wicked, but manageable if social fragmentation is minimized. Consider, for example, the "problem" of sending a person to the moon. The complexity of such a problem is clearly enormous—even though the physics may be relatively straightforward. As we know, the complexity of a manned moon landing was quickly overcome by a relatively cohesive and committed society. Wickedness, in this case, was wonderfully tamed.

Unfortunately, cohesiveness and commitment appear to be absent in our struggle to appropriately address global climate change. Cohesiveness is replaced by the fragmentation that occurs when the scientific method is replaced by fear and distrust of diverse arguments and points of view. Commitment to understanding and solution makes way for denial of the many indicators that things are going terribly wrong. In the end, as Conklin points out, "Because of the complexity surrounding wicked problems, solving such a problem is fundamentally a social process. Having a few brilliant people or the latest project management technology is no longer sufficient."

If the solution to any emergent complex environmental problem—whether or not you subscribe to a major anthropogenic component of those problems—is indeed a social process, then the first step in that process is surely exactly the same as the first step in any human problem solving endeavor—as well as the first step in a complete description of the Scientific Method—which is to engage, analyze, and communicate a thorough act of observation. And any act of environmental observation is increasingly sobering. The figure below, for example, illustrates a range of observations that, for better or worse, are perturbing every known planetary feedback system.

Faced with these observations, along with many other complementary ones, we appear to have three choices: 1) lock down the problem definition and focus, for example, on carbon emissions, 2) deny that we have a problem, or 3) minimize social fragmentation and commit to a sustainable future. As we engage each other in the continuing debate about the nature of our path forward, we might consider carefully the words of Theodore Roosevelt: "To waste, to destroy our natural resources, to skin and exhaust the land instead of using it so as to increase its usefulness, will result in undermining in the days of our children the very prosperity which we ought by right to hand down to them amplified and developed."[5] This idea, I believe, could serve as the vehicle that separates us from our fragmented past and delivers us to a sustainable future.

Figure 1. Indicators of Environmental Change[4]

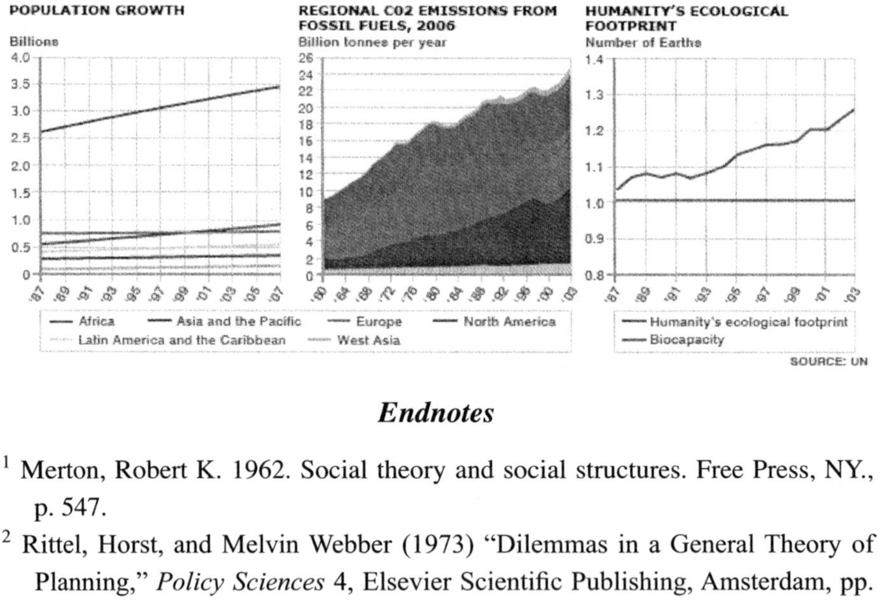

POPULATION GROWTH
Billions

REGIONAL C02 EMISSIONS FROM FOSSIL FUELS, 2006
Billion tonnes per year

HUMANITY'S ECOLOGICAL FOOTPRINT
Number of Earths

— Africa — Asia and the Pacific — Europe — North America
Latin America and the Caribbean — West Asia

— Humanity's ecological footprint
— Biocapacity

SOURCE: UN

Endnotes

[1] Merton, Robert K. 1962. Social theory and social structures. Free Press, NY., p. 547.

[2] Rittel, Horst, and Melvin Webber (1973) "Dilemmas in a General Theory of Planning," *Policy Sciences* 4, Elsevier Scientific Publishing, Amsterdam, pp. 155-159. Also Reprint No. 86, The Institute of Urban and Regional Development, University of California, Berkeley, California.

[3] Conklin, Jeff; "Wicked Problems & Social Complexity," Chapter 1 of *Dialogue Mapping: Building Shared Understanding of Wicked Problems*, Wiley, October 2006.

[4] Copyright © 2007, United Nations Environment Programme.

[5] Theodore Roosevelt, seventh annual message, 3 December 1907.

GOVERNMENTS AND CLIMATE CHANGE ISSUES:
A FLAWED CONSENSUS

David Henderson*

Governments are mishandling climate change issues. Both the basis and the content of official policies are open to serious question. As to the former, too much reliance is placed on the large-scale established process of review and inquiry which is conducted through the agency of the Intergovernmental Panel on Climate Change. This process, which is wrongly taken to be objective and authoritative, has been made the point of departure for over-presumptive conclusions which are biased towards alarm, in the mistaken belief that "the science" is "settled." Governments should take prompt steps to strengthen the basis for policy, by ensuring that they and their citizens are more fully and more objectively informed and advised. This implies both improving the IPCC process and going beyond it.

I believe that governments across the world, most notably the governments of the Organization for Economic Cooperation and Development (OECD) member countries, are mishandling climate change issues. The mishandling has two related aspects. First, actual policies to curb "greenhouse gas" emissions too often take the form of costly specific regulations, rather than a general price-based incentive such as a carbon tax. More fundamentally, there is good reason to question the basis and rationale of policy—the arguments, beliefs, and presumptions which have led so many governments to take action. It is on this latter aspect that I focus this paper. By way of setting the scene, I begin by outlining the present state of affairs and the events leading up to it.

A World-Wide Official Consensus

In relation to climate change issues, there exists a broad and well-established official consensus. With few exceptions, governments are firmly committed to the view that anthropogenic global warming constitutes a

---

* This paper, prepared November 28, 2007, draws at various points, without specific attribution, on an article of mine which was published in *World Economics*, vol. 8, no. 2 (April-June 2007).

serious problem which requires official action at both national and international levels. A recent high-level restatement of this consensus position is contained in the Declaration issued at the close of the G-8 Summit meeting in Heiligendamm, Germany, in June 2007. In paragraph 49 of the Declaration, the G-8 leaders said that "global greenhouse emissions must stop rising, followed by substantial global emission reductions." They thus reaffirmed the case for what are often described as "mitigation" policies, designed to curb emissions.

In pretty well every democratic country, this official consensus is not at all a matter of political controversy: to the contrary, it enjoys general cross-party support. In the world as a whole, I can think of only one political leader who is a convinced and open dissenter—namely, the President of the Czech Republic, Vaclav Klaus. Governments generally, and opposition parties too where they exist, have determined that policies designed to curb emissions are called for and that the existing array of measures needs to be extended and reinforced.

This official, multi-partisan consensus is not new. Climate change issues, and in particular the extent and possible consequences of anthropogenic global warming, have been on the international agenda for 20 years or more; and it is now over 15 years since governments decided, collectively and almost unanimously, that determined steps should be taken to deal with what they agreed was a major problem. The decisive collective commitment was made in 1992, through the United Nations Framework Convention on Climate Change (UNFCCC), which almost all countries have ratified. The Convention specifies that its "ultimate objective" is

To achieve . . . stabilization of greenhouse gas emissions in the atmosphere at a level that would prevent dangerous anthropogenic interference with the climate system.

Precisely this form of words is repeated in the Heiligendamm G-8 Summit Declaration.

Since 1992, many governments have acted, at state and provincial as well as national levels and collectively within the European Union, through what is now a wide range of measures and programs, to curb emissions of (so-called) "greenhouse gases." On the international scene, through the Kyoto Protocol (1997), "Annex I" countries have undertaken to meet specific targets for emissions reductions. It is true that these Kyoto-based commitments are viewed by many as relatively unambitious, or as a first step only, and that in almost all the countries concerned they seem unlikely to be met. But the accepted direction of policy remains clear and unquestioned; and both nationally and internationally, new and far-reaching measures to curb emissions are under consideration or in prospect.

In taking this line, governments have met with widespread and increasing public approval. Prominent among the unofficial sources of support are media commentators on environmental and scientific issues, scientific bodies, environmental non-governmental organizations (NGOs), and, increasingly, large business enterprises. Let me add that there is widespread support for the official consensus position among economists, as evidenced for example in the Stern Review (2006) on *The Economics of Climate Change* and some of the reactions to it. As usual, however our profession is not of one mind.

The Basis for Consensus

What was it that persuaded governments across the world, 15 or more years ago, to take the possible dangers of anthropogenic global warming so seriously, and what is it that has caused them to maintain and even intensify their concerns? I think the answer is straightforward. From the start the main influence was, as it still is, *the scientific advice provided to them.* That advice can and does come from many sources; but the main single channel for it, indeed the only channel of advice for governments *collectively*, has been the series of massive and wide-ranging Assessment Reports produced by the Intergovernmental Panel on Climate Change, the IPCC.

The IPCC was established by the governments of the world in 1988 as the joint subsidiary of two UN agencies, the World Meteorological Organisation (WMO) and the United Nations Environment Program (UNEP). Its first Assessment Report, which appeared in 1990, formed a basis and point of departure for the negotiations that led up to the drafting of the Framework Convention. Since then, the Panel has published three further reports of the same kind. The fourth and latest of these, referred to for short as AR4, was completed and published in the course of 2007. As with earlier reports, it chiefly comprises the separate volumes issued by each of the Panel's three Working Groups. Among them these three volumes, each with its own Summary for Policymakers, come to around 3,000 pages, and some 2,500 experts—authors, contributors and reviewers—were directly involved in preparing them: I refer to this small army of participants as the IPCC *expert network*. AR4 was finally rounded off with an overall Synthesis Report.

These IPCC Assessment Reports are far-reaching — indeed, they are uniquely broad in scope. They deal with the whole range of issues relating to climate change, including economic as well as scientific and technical aspects. In producing them, the Panel has brought together teams of specialists drawn from across the world and put in place ordered procedures for directing and reviewing their work and arriving at agreed final texts. It has secured for the reports and their conclusions the acceptance of its many and diverse member governments; and in consequence, it has helped to guide the thinking of those governments.

The IPCC does not itself undertake or commission research: the Assessment Reports review and draw on the already published work of others. Most of this work is financed by governments, and these governments thus have their own direct sources of information and advice — their thinking and actions do not necessarily depend on what the Assessment Reports have said. All the same, the IPCC's work carries substantial weight, with public opinion as well as the Panel's member governments, because of its wide-ranging coverage, its extensive and ordered scientific participation, and the fact that it alone is designed to serve and inform the world as a whole.

An explicit tribute to the work of the Panel is paid in the G-8 Summit Declaration. The words that I quoted from the Declaration, at the beginning of my presentation, comprised only part of the sentence in question. The full sentence reads, with italics added:

> Taking into account the *scientific knowledge* as represented in the recent IPCC reports, global greenhouse emissions must stop rising, followed by substantial global emission reductions.[1]

More recently the work of the Panel has received further and conspicuous international recognition through the award of the 2007 Nobel Peace Prize, which it shared with Al Gore. The citation for the award says approvingly that the Panel "has created an ever-broader informed consensus about the connection between human activities and global warming," and actually this form of words does not do justice to the full range of topics that the IPCC covers.

On the basis of the three Assessment Reports that the IPCC has produced

64

since the Framework Convention was signed, and of AR4 in particular, governments have certainly no reason to question the consensus position that they adopted more than 15 years ago. To the contrary, all three reports have served to confirm and strengthen that position.

So much for the official and widely accepted consensus and its basis. Given this background, you might well want to ask me for an explanation. How is it that I, as an outsider and a non-participant, an economist and not a scientist, have come to question the considered stance which so many governments have continued to take, on the basis of the scientific advice they have been given and with substantial and increasing public support, including support from scientific bodies and, as I have noted, from many of my fellow economists? What reasons do I have for holding that the advice, and the conclusions drawn from it, is open to serious question?

In responding, I will focus chiefly on the role and work of the IPCC. The Assessment Reports are seen as giving expression to a worldwide scientific consensus, based on an informed and objective professional evaluation, and therefore providing a sound basis for policy. Let me explain why I have come to question this picture.[2]

Panel and Process

Why do governments, and outsiders too, place so much trust in the IPCC's role and work? I think that the trust largely results from the wide and structured expert participation that the IPCC process ensures. People visualize an array of technically competent persons whose knowledge and wisdom are effectively brought to bear through an independent, objective and thoroughly professional scientific inquiry. Indeed, many observers identify the Panel with the network, as though well-qualified and disinterested experts were the only people involved.[3] The reality is both more complex and less reassuring.

A basic distinction has to be made between the IPCC as such, that is to say the *Panel*, and the IPCC *process*. The two are not the same, and the process involves three quite distinct groups of participants.

The first of these groups comprises the *Panel* itself, which controls the preparation of the reports, along with its two subsidiary bodies. The Panel effectively comprises those officials whom governments choose to send to Panel meetings. My impression is that, generally speaking, these are not high-ranking persons. They include scientists as well as laymen. Numbers are not fixed, but a typical Panel meeting might involve some 300-400 participants. Working directly for the Panel is the IPCC *Secretariat*, though this is a small group whose functions are mainly of a routine administrative kind. A more influential body is the 28-strong IPCC *Bureau*, comprising high-level experts in various disciplines from across the world, chosen by the Panel. The Bureau acts in a managing and coordinating role under the Panel's broad direction.

The second group is made up of the 2,500-strong *expert network*, the persons who put together the draft Assessment Reports. This network is separate and distinct from the Panel itself. There is little or no overlap between the two bodies.

Last but far from least, there are the government departments and agencies which the Panel reports to: it is here, and not in the Panel itself, that the ultimate "policymakers" are to be found. The relevant political leaders and senior officials within these departments and agencies largely make up what I call the *environmental policy milieu*. This milieu also comprises leading non-official members of the IPCC Bureau, past as well as current, and together with the most influential members of the Panel itself, these persons make up what may be termed the informal *directing circle* of the IPCC. In turn, the directing circle, together with a substantial number of prominent and like-minded expert participants in the reporting process, can be seen as making up an informal *IPCC milieu*.

Policy Commitment

The IPCC as such has been formally instructed by its member governments, in the "principles governing IPCC work," that its reports "should be neutral with respect to policy." However, this instruction must be interpreted as referring specifically and exclusively to the contribution made by the expert network through the reporting process. It does not, *and could not*, apply to the other two participating groups. The official Panel members, as also the policy milieu which they report to, are almost without exception far from neutral: they are committed, *inevitably and rightly*, to the objective of curbing emissions, as a means to combating climate change, which their governments agreed on when they ratified the Framework Convention; and in many cases they are likewise committed to the kinds of policies that their governments have adopted in pursuit of that objective. As officials, they are bound by what their governments have decided. That is

66

the context within which the three successive IPCC Assessment Reports prepared since 1992 have been put together by the network and reviewed by member governments. The clients and patrons of the expert network, with few exceptions, take it as given that anthropogenic global warming is a serious problem which demands, and has rightly been accorded, both national and international action.

It is against this background, of a Panel and a controlling policy milieu that are not and could not be "policy-neutral," that some basic features of the expert reporting process have to be borne in mind:

- The choice of lead authors for the Assessment Reports largely rests with the already committed member governments, since lists that they provide form the starting point for the selection process;
- Complete draft texts of the Working Group reports go to these governments for comment and review; and
- It is governments, as represented in the Panel, that sign off on the final versions of the Assessment Reports and which amend the draft Summaries for Policymakers and the final Synthesis Report, before they approve these also for publication.

The fact is that departments and agencies which are not—and cannot be—uncommitted in relation to climate change issues are deeply involved, from start to finish, in the preparation of the Assessment Reports.

Does this fact in itself put in question the expert reporting process and the Assessment Reports? As a former official myself, I would say: No, not necessarily. Policy commitment on the part of member governments could in principle go together with a resolve on the part of the policy milieu, and of the Panel which they appoint and control, to ensure that the reporting process is open, thorough, objective and policy neutral. This indeed is what governments believe, or at least maintain, is the state of affairs that they have created; and I think many outside persons believe or presume the same. In this generally accepted picture of the IPCC process, an invisible Chinese wall separates the committed patrons and clients of the reporting process from the array of disinterested scientists, policy neutral in their expert capacity, who take part in it.

I have come to believe that this picture is not accurate, and that the expert reporting process is flawed. Despite the numbers of persons involved and the lengthy formal review procedures, the preparation of the IPCC Assessment Reports is far from being a model of rigor, inclusiveness, and impartiality. In my view, the flaws in the process can be largely accounted for by a pervasive bias on the part of the people and organizations that direct and control it. I shall comment first on some flaws and then on the forms and sources of bias.

Errors, Omissions, and Bias

Despite the numbers involved, the expert process has not ensured appropriately broad professional involvement. A case in point is the treatment of statistical issues. A leading American statistician, Edward Wegman, has noted that

> the atmospheric science community, while heavily using statistical methods, is remarkably disconnected from the mainstream community of statisticians in a way, for example, that is not true of the medical and pharmaceutical communities.

As for economics, Ross McKitrick, in written evidence to the House of Lords Select Committee on Economic Affairs, argued that after the Second Assessment Report, which appeared in 1995, "the IPCC could no longer claim to have the participation of mainstream professional economists." I think that the subsequent list of AR4 participants lends support to this view.[4]

In relation to economic issues, a specific weakness in some IPCC documents has been the use of invalid cross-country comparisons of output (real GDP), based on exchange rates rather than purchasing power parity (PPP) estimates; such comparative figures give a distorted picture of the world economy and the course of economic change. Some further misleading observations on this central topic are to be found in AR4.[5]

A basic general weakness in the reporting process is the uncritical reliance on peer review as a qualifying criterion for published work to be taken into account. Peer review provides no safeguard against dubious assumptions, arguments, and conclusions if the peers are largely drawn from the same restricted professional milieu. What is more, the peer review process as such may be insufficiently rigorous. In particular, in cases where research has involved the assemblage and processing of large data sets, peer review does not guarantee due disclosure of sources, methods, and procedures so that results can be replicated by others.

Failures of disclosure, of a kind that some leading academic journals would not tolerate and which would not be permitted in business prospectuses, have characterized published work that the IPCC has drawn on. The most notable case is that of the temperature reconstructions which entered into what became known as the "hockey stick" study. This piece of work, which was prominently featured and drawn on in the IPCC's Third Assessment Report and afterwards, formed the basis for a striking and much-publicized claim that in the Northern Hemisphere the 1990s had been the warmest decade of the millennium and 1998 the warmest single year. Probably no single piece of alleged evidence relating to climate change has been so frequently cited and influential. The authors concerned failed (and later declined, until strong pressures were eventually brought to bear) to make due disclosure, and neither the publishing journals nor the IPCC required them to do so. Resistance to disclosure was eventually overcome only through a U.S. congressional committee investigation.

Further issues of disclosure, and of the treatment of evidence, have arisen in relation not only to subsequent temperature reconstructions but also to the instrument-based temperature series that the IPCC reports have relied on. In this latter context, eventual release of pertinent information has recently been secured only by bringing to bear British freedom of information legislation.

In these various cases, from the "hockey stick" study onward, exposure of the problem, and the pressures for due disclosure, have come largely from private individuals: so far as I know, not a single government department or international agency has faced up to the issues. Prominent among the individuals concerned have been two Canadian authors, Stephen McIntyre and Ross McKitrick. Both separately and in joint writings, they have made an outstanding contribution to public debate.[6]

A related issue which has recently come into prominence is the treatment by IPCC lead authors, during the review process, of critical comments and suggestions for changes in their drafts. Here again, it has been necessary to use freedom of information laws to break down official resistance to publication of the relevant exchanges, and the objectivity of some authors and of the review process has been put in question.[7]

The handling of these various disclosure lapses by the IPCC's directing circle reflects no credit on those involved: they have failed to acknowledge the problem and take appropriate action. In the relevant sections of AR4 the issue is evaded, while a misleading picture is presented of the various writings on the subject of temperature reconstructions. Here as elsewhere, the response of the IPCC milieu to informed criticism has been inadequate, evasive or dismissive.[8]

In the "principles governing IPCC work," laid down by governments and already quoted above, it is specified that the work of the Panel should be "open and transparent." But one cannot apply these terms to a process in which key participants fail to disclose information that should from the start have been available in full, where such disclosure failures are condoned by those who control and direct the process, and where the information is eventually made available only through the agency of a congressional inquiry and resort to freedom of information laws.

I have now come to think — and the thought had not formed in my mind when I first became involved with climate change issues, more by accident than design, five years ago — that the IPCC process, viewed as a whole and including the expert reporting process, is not professionally up to the mark. I think that the main reason for this deficiency is a strong and continuing element of bias that has always been present within both the environmental policy milieu and the IPCC directing circle. This ingrained bias goes beyond the formal commitment of the officials concerned to the established post-1992 intergovernmental consensus.

Instances and Forms of Bias

One aspect of prevailing official bias emerges indirectly from the public debate on climate change issues. Across the world, the treatment of these issues by environmental and scientific journalists and commentators is overwhelmingly one-sided and sensationalist: studies and results that are unalarming are typically played down or disregarded, while the gaps in knowledge and the huge uncertainties which still loom large in climate science are passed over. A conspicuous recent case in point, both in itself and in its reception by the media, is the Al Gore film and book, *An Inconvenient Truth* (2006). This pervasive one-sidedness on the part of so many commentators and media outlets is in itself worrying; but even more so, in my mind, is the fact that leading figures and organizations connected with the IPCC process, including government departments and international agencies, do little to ensure that a more balanced picture is presented. It is characteristic of the official environmental policy milieu that some govern-

70

ments, including my own, have chosen to distribute *An Inconvenient Truth* to schools as an officially recommended and reliable source.

More direct evidence of bias can be seen from the kinds of statements about climate change issues that are freely made in many countries by leading public figures. The tone of these is distinctly alarm-prone. As in the case of Al Gore and many other commentators, it is taken as established beyond question that humankind is placing the planet under dire threat, that in consequence further drastic measures of mitigation are urgently required, and that such measures would be effective in determining climate outcomes.

Disaster Scenarios

Here are some summit-level instances of this way of thinking, which I call the *heightened milieu consensus*:

- Tony Blair, then still Prime Minister of the U.K., commenting a year ago on the Stern Review on the economics of climate change, said "What is not in doubt is that the scientific evidence of global warming caused by greenhouse gas emissions is now overwhelming . . . [and] . . . that if the science is right, the consequences for our planet are literally disastrous."
- Blair and the Dutch prime minister, in a joint letter of October 2006 to other EU leaders, wrote "We have a window of only 10–15 years to take the steps we need to avoid crossing a catastrophic tipping point."
- Stephen Harper, Prime Minister of Canada, in a speech earlier in 2007, described "climate change" as "perhaps the biggest threat to confront the future of humanity today."
- President Nicolas Sarkozy of France, in some remarks in May 2007, shortly before his election to office, declared "What is at stake is the fate of humanity as a whole."

These assertions, though they are in tune with much public thinking, go beyond the sober language of the G-8 Summit Declaration, and they are not drawn directly from IPCC Assessment Reports. They are bold extrapolations from the Reports, with a clear presumptive element.

Interestingly, such statements have been criticized by a leading British climate scientist, Professor Mike Hulme, as forms of what he called "a

71

discourse of catastrophe [which] is a political and rhetorical device." Referring to the second of the quotations above from Tony Blair, he described our then-Prime Minister as among "recent examples of the catastrophists" (Hulme 2007).

Now while it can be argued that Blair deserved to have this label attached to him for such remarks, it is not with him that the chief responsibility for them rests. He and his Dutch cosignatory, as also Harper and Sarkozy in the quotations above, almost certainly did not write their own speeches. What they said was presumably sanctioned, if not actually drafted, by their scientific and environmental advisers and by the departments in which those people work; and had it not been so sanctioned, those advisers could have ensured that future public statements would take a more measured and qualified tone.

In the statement that I just quoted, Professor Hulme draws a contrast between catastrophists and climate scientists. As I see it, however, there is no clear dividing line between the two. It is climate scientists who write, or lend tacit approval to, the catastrophist scripts of leading lay figures, and who in some prominent cases have made similar pronouncements of their own.

When it comes to leading officials with an advisory role, it is not difficult to find advocates of the heightened consensus. Within the British government machine, the Chief Scientific Adviser, Sir David King (2004), has taken the position that "climate change is the most severe problem that we are facing today—more serious even than the threat of terrorism." As to leading figures within the IPCC directing circle, here are three recent instances. The following are public statements made in February 2007, following the publication of the report of Working Group I, which forms the first volume of AR4:

- Dr. R.K. Pachauri, the Chairman of the IPCC, and hence of the IPCC Bureau: "I hope this report will shock people [and] governments into taking more serious action."
- Achim Steiner, the Director-General of the UNEP: "In the light of the report's findings, it would be 'irresponsible' to resist or seek to delay actions on mandatory emissions cuts."[9]
- Yvo de Boer, Executive Secretary of the UN Framework Convention: "The findings leave no doubt as to the dangers that mankind is facing and must be acted on without delay."

These are strong assertions. In none of them was the wording taken directly from the report in question: these eminent persons were going beyond the Working Group I text to draw their own confident and unqualified personal conclusions as to the lessons for policy. While they were fully entitled to form and air such opinions, their statements were not just summaries of "the science," nor of course were they "policy neutral."

In speaking as they did on this occasion, these three leading figures were conforming to an established pattern. From the earliest days, most if not all of those directing the IPCC process, within governments and outside, have shared the conviction that anthropogenic global warming presents a threat, to humanity and the planet, which demands prompt and far-reaching action by governments: and had this *not* been the case, and *known* to be the case, *they would not have attained their leading positions within the process.* To take only the three current examples just quoted: Pachauri, Steiner, and de Boer would not have sought their respective posts, nor would they have been seen by UN agencies and member governments as eligible to hold them, had they not been identified as fully committed to "heightened consensus" views. The same has been true throughout of the IPCC Bureau and other leading figures within the process. The process is run today, as it has been from the start, by true believers. This accounts for the readiness of many of those concerned to make strong public pronouncements of the kind quoted above, which go beyond the more nuanced language of the Assessment Reports; to turn an unseeing eye to the disclosure failures and other professional flaws in the reporting process; and to view with equanimity or approval the lack of balance that characterizes public debate.

An Increasingly Conformist Network?

It would of course be wrong to presume that the attitudes and beliefs that I have just described, which characterize the environmental policy milieu and the IPCC directing circle, are shared by of all those who make up the expert network. However, my impression is that over time that network, while growing in numbers (so that the stock of peer reviewers has expanded *pari passu*), has become increasingly influenced, if not dominated, by subscribers to the official consensus, often in its heightened form. It has become more difficult for independent outsiders, who do not share accepted beliefs and presumptions of the IPCC directing circle and of the Panel's parent bodies and sponsoring government departments and agencies—which, it has to be remembered, provide the overwhelming bulk of research funding in this area—to contribute usefully to the reporting process. For this and other reasons, some nonconforming experts have either declined to become involved with the process or have later withdrawn from it. The network has thus become more numerous but less inclusive. At the same time, it may have become harder for younger scientists, with careers still to make, not to fall in with received majority opinion which is both officially sponsored and strongly held. In evidence to the House of Lords Select Committee on Economic Affairs (Vol. II, p. 233), David Holland wrote, admittedly as an outsider: "If I were beginning my career I cannot imagine that I could make a living in climate science without accepting the current consensus." In both scientific circles and the reporting process, some dissenters have held aloof, while others have been gradually sidelined or eased out. This evolution forms part of the background against which the professional lapses noted above are to be viewed and judged.

The Influence of Global Salvationism

Some history is relevant here. Within the environmental policy milieu, together with a range of outside allies, there is a generic bias which goes a long way back and extends well beyond issues relating to climate change. Over a period of some 40 years, and increasingly over time, departments and agencies concerned both with the environment and with the economic problems of poor countries have typically adhered to the set of beliefs and presumptions which make up what I have termed *global salvationism*.[10]

In the salvationist picture of reality, two elements are combined. One is an unrelentingly somber picture of recent trends, the present state of the world (or "the planet"), and prospects for the future unless governments involve themselves more closely, and with immediate effect, in the management and control of events. Within this picture, environmental issues are treated almost exclusively in terms of problems, dangers, and potential or even imminent disasters, with the presumed harmful effects of economic growth as one reason for concern. The second element is a conviction that known effective remedies exist for the various ills and threats thus identified: "solutions" are at hand, given wise collective resolves and prompt action by governments and "the international community." Global salvationism thus combines dark visions and alarming diagnoses with confidently radical collectivist prescriptions for the world.[11]

74

The essence of a widely accepted diagnosis is conveyed by the following quotation, which comes from a mid-1970s Club of Rome study:

Two gaps, steadily widening, appear to be at the heart of mankind's present crises: the gap between man and nature, and the gap between "North" and "South," rich and poor. Both gaps must be narrowed if world-shattering catastrophes are to be avoided. . . .[12]

By the end of the 1970s a broad milieu salvationist consensus had become well entrenched, not only in national capitals but also in a range of UN agencies, and with widespread public support. During the 1980s this view of the world found expression in two widely read and influential reports, each produced by a specially convened international group of eminent persons. The first of these was the *Brandt Report* of 1980, and the second, more influential, *Brundtland Report* of 1987.[13] Included in the latter was a section on the possible dangers from anthropogenic global warming, which was described (p. 34) as "a threat to life-support systems," and from that time on a belief in the reality of such a threat came to be an integral part of global salvationist doctrine. The *Brundtland Report* led on to the December 1989 resolution of the UN General Assembly, which authorized what became the 1992 UN Conference on Environment and Development (the Rio "Earth Summit"), at which the Framework Convention was formally adopted.

Rio and After

Climate change was not the only item on the Earth Summit agenda, and the Framework Convention was not the only document on which agreement was reached.

Within the Rio documents and resolutions, the familiar UN-style dark salvationist message was neither qualified nor watered down. Chief among the many documents prepared for the Summit was a 600-page proposed action program called Agenda 21, which the conference actually adopted with some amendments. The preamble to this text opens as follows:

Humanity stands at a defining moment in history. We are confronted with a perpetuation of disparities within and between nations, a worsening of poverty, hunger, ill health, and illiteracy, and the continued deterioration of the ecosystems on which we depend for our well-being.

The participants who signed up to these misleading salvationist assertions included George Bush (Senior), who was then President of the United States, John Major as the British Prime Minister, and Bob Hawke as the Prime Minister of Australia.

The proposed remedies for the stark situation thus depicted in Agenda 21 were to be given effect through "a new global partnership for sustainable development." This wording was significant. The Earth Summit marked the general endorsement by governments of the principle of "sustainable development" as a basis for policy.

Such was the salvationist diagnosis and prescription which all the participating governments at Rio proved ready to endorse, many of them at the highest political level. Into these Summit proceedings the IPCC's contribution entered, not as a separate stand-alone scientific exercise, but rather as a powerful new reinforcing element in the already existing and widely accepted salvationist picture of the world. Because of this reinforcement, global salvationism gained further and influential endorsement in official circles: it became the creed, not just of environmental departments and agencies and their supporting NGOs, but of governments as a whole. In pointing to new threats and new "solutions," the issue of global warming served then, as it still serves today, to give expression to, and extend, established alarm-prone convictions.

Of course, this historical link with questionable salvationist beliefs can be seen as no more than coincidental: in itself, such an association does not put in doubt the findings of climate scientists or the competence and objectivity of the IPCC expert network and the reporting process. It is possible to accept the present official consensus on climate change issues, and the IPCC's latest Summaries for Policymakers, without signing up to the distorted picture of the world given in Agenda 21 and its successors. Indeed, it is not difficult to find strong critics of global salvationist pessimism who nonetheless accept that anthropogenic global warming is both real and a cause for concern: a prominent example is Bjørn Lomborg (2001 and 2007). However, I believe that the close relationship between the IPCC milieu and its sponsoring departments and agencies, together with the already ingrained salvationist propensities of both, have from the start, and increasingly over time, put in question the objectivity of the IPCC process and hence its claims to authority. The professional advice which governments continue to rely on has been, and still is, suffused with bias.

76

Summing Up

In relation to climate change issues, governments have locked themselves into a set of procedures, and an associated way of thinking—in short, a *framework*—which both reflects and yields over-presumptive conclusions which are biased towards alarm. These conclusions form the basis both of current policies, which incidentally raise problems of their own, and of proposals to take those policies considerably further. They go beyond the bounds of professional consensus, they take as their prime source the results of a flawed process, and they represent a dubious extension of those results.

Even if the IPCC process were beyond challenge, it is imprudent for governments to place such heavy reliance, in matters of extraordinary complexity where huge uncertainties remain, on this particular source of information, analysis and advice. In fact, the process is flawed, and this puts in doubt the accepted basis of the established official consensus.

In relation to climate change, and not only in the context of the IPCC, there is a clear present need to build up a sounder basis for reviewing and assessing the issues. Governments should try to ensure that they and their citizens are more fully and more objectively informed and advised.

Post-Script: What Can Be Done?

In considering how the present situation might be improved, the main focus has to be on governments. It is they that fund major programs and decide policies, while only they can reform the process which they have created and over which they have full control. In that connection, let me put forth just one central argument and point to some of its implications.

My argument is this: So long as the handling of climate change issues is left almost entirely to environmental departments and agencies, there is little or no prospect of reform. A necessary condition for change, albeit not a sufficient condition, is that other departments of state should become effectively involved.

In particular, since the economic stakes could be high, a responsibility here rests on the economic departments of state—treasuries, ministries of finance and economics, and, in the U.S., the Council of Economic Advisers. I am myself a former Treasury official, and much later, as Head of what was then the Economics and Statistics Department in the OECD Secretariat, I had close dealings over a number of years with economics and finance ministries in OECD member countries. I have been surprised by the failure of these ministries to come to grips with climate change issues, their uncritical acceptance of the results of a process of inquiry which is so obviously biased and flawed, and their lack of attention to the criticisms of that process which have been voiced by independent outsiders—criticisms which, as I think, they ought to have been making themselves.

Such a conclusion points to official action on four related fronts:

- First, governments could improve the IPCC process by making it more professionally watertight. For a start, they should insist on true and full disclosure as a precondition for published work to be taken into account in the review process.
- Second, they should no longer presume or aim at consensus. Rather, they should see to it that, both within the IPCC reporting process and more broadly, serious differences of professional opinion are aired.
- Third, they should consider developing sources of information and advice that are independent of the IPCC process, thus bringing to an end the Panel's virtual monopoly status as a source of collective advice.
- Fourth, they should broaden the basis of official participation, so that it goes beyond the existing well-entrenched environmental policy milieu.

Not all of these lines of action require international agreement: much could be done by individual governments acting on their own account. If even one or two influential governments were to question their current presumptions, and act accordingly, this could change the whole situation.

References

Ian Castles and David Henderson (2005), "International Comparisons of GDP: Issues of theory and practice," *World Economics*, vol. 6, no. 1.

Robert Ehrlich (2003), *Eight Preposterous Propositions: From the Genetics of Homosexuality to the Benefits of Global Warming*, Princeton, NJ: Princeton University Press.

P. D. [David] Henderson (1980), "Survival, Development and the Report of the Brandt Commission," *The World Economy*, vol. 3, no. 1.

David Henderson (2004), *The Role of Business in the Modern World: Progress, Pressures, and Prospects for the Market Economy*, published by Institute of Economic Affairs (London, UK), New Zealand Business Roundtable (Wellington), and Competitive Enterprise Institute (Washington, DC).

David Henderson (2005), "SRES, IPCC, and the Treatment of Economic Issues: What Has Emerged?" *Energy and Environment*, vol. 16, no. 3/4.

David Henderson (2007), "Governments and Climate Change Issues: The case for rethinking," *World Economics,* vol. 8, no. 2 (April-June).

David Holland (2007), "Bias and Concealment in the IPCC Process: The 'Hockey-Stick' Affair and Its Implications," *Energy and Environment*, vol. 18, no. 7/8.

House of Lords, Select Committee on Economic Affairs (2005), The Economics of Climate Change: Vol. I, Report; Vol. II, Evidence. 2nd Report, 20052006 session.

House of Lords, Select Committee on Economic Affairs (2006), Government Response to the Economics of Climate Change, HL Paper 71. 3rd Report, 2005-2006 session.

Mike Hulme (2006), "Chaotic world of climate change," broadcast as a *Viewpoint* on BBC News, November 2006, and available on the author's website.

Independent Commission on International Development Issues (the Brandt Report) (1980), North South: A Programme for Survival, Pan Books. See summary at http://www.stwr.net/content/view/43/83/.

David A. King (2004), "Climate Change Science: Adapt, Mitigate, or Ignore?" *Science*, vol. 303, no. 5655.

Bjorn Lomborg (2001), The Skeptical Environmentalist, Cambridge, UK: Cambridge University Press.

Bjorn Lomborg (2007), Cool It: The Skeptical Environmentalist's Guide to Global Warming, New York, NY: Alfred A. Knopf.

Stephen McIntyre and Ross McKitrick (2003), "Corrections to the Mann et. al. (1998): Proxy Data Base and Northern Hemisphere Average Temperature Series," *Energy and Environment*, vol. 14, no. 6.

Stephen McIntyre and Ross McKitrick (2005), "Hockey Sticks, Principal Components and Spurious Significance," *Geophysical Research Letters*, vol. 32, no. 3.

Stephen McIntyre and Ross McKitrick (2005), "The M&M Critique of the

MBH98 Northern Hemisphere Climate Index: Update and Implications," *Energy and Environment*, vol. 16, no. 1.

Ross McKitrick (2006), "Bringing balance, disclosure, and due diligence into science-based policymaking," in Jene Porter, ed., *Public Science in Liberal Democracy: The Challenge to Science and Democracy*, Toronto, Ontario: University of Toronto Press.

Mihajlo Mesarovic and Eduard Pestel (1975), *Mankind at the Turning Point: The Second Report to the Club of Rome*, London, UK: Hutchinson.

Nicholas Stern and others (2006), The Economics of Climate Change: The Stern Review, Cambridge, UK: Cambridge University Press.

World Commission on Environment and Development (the Brundtland Report) (1987), Our Common Future, London, UK: Oxford University Press.

John Zillman (2007), "Some Observations on the IPCC Assessment Process, 1988-2007," *Energy and Environment*, vol. 18, no. 7/8.

Endnotes

1 Had I been one of the official "Sherpas" involved in preparing the draft Declaration, I would have argued for using the term "judgments" here, rather than "knowledge," which I think goes too far.

2 The IPCC's role and work form the subject of a group of articles, including one of mine, in the journal *Energy and Environment*, vol. 18, no. 7/8 (2007). The authors include John Zillman, who was a leading participant in the IPCC's work from the earliest stages. His article (Zillman 2007) offers a much more favorable assessment of the IPCC process than mine.

3 Among leading scientists, an example is Robert Ehrlich, a professor at Yale University. He has described the IPCC as "a respected international group of hundreds of scientists" and as "comprised of scientists from 99 countries" (Ehrlich, [2005], pp. 138, 169). As will be seen, this is not a correct statement.

4 Professor Wegman's observation was made in one of his responses to questions put to him in the course of an inquiry carried out by a committee of the U.S. House of Representatives. McKitrick's evidence to the Select Committee is in the Committee's report, vol. 2 (2005), pp. 262-263.

5 From late 2002 on, Ian Castles and I jointly put forward a critique of some leading aspects of the IPCC's economic work, while authors involved

in that work contested our criticisms. Following these exchanges, we published in 2005 a joint paper on international comparisons of GDP, and I reviewed and carried further the whole debate in a later article (Henderson 2005).

[6] Some key publications of theirs are noted in the list of references above.

[7] The whole complex of issues covered in these last few paragraphs—of non-disclosure, non-response, selective coverage, and bias within the reporting process—is treated at length in a recently published paper by David Holland (2007).

[8] A characteristic instance was the British government's dismissive official response (2006) to the 2005 report of the House of Lords Select Committee on Economic Affairs on the economics of climate change.

[9] This and the following quotation are taken from a report in the Financial Times, February 3, 2007.

[10] The content, history, and implications of global salvationism form the main theme of Chapter 4 of Henderson (2004).

[11] A prominent feature of the dark salvationist picture of reality has been much-overstated measures of the gap between rich and poor countries, derived from invalid exchange-rate-based, rather than Purchasing Power Parity (PPP)-based, comparisons of GDP per head.

[12] Mihajlo Mesarovic and Eduard Pestel, *Mankind at the Turning Point: The Second Report to the Club of Rome*, London, UK: Hutchinson (1975), p. ix.

[13] At the time, I published a review article on the *Brandt Report* (Henderson 1980), where my final assessment of the document as a whole was that "the view of the world on which it rests is false."

RESPONSE TO HENDERSON ARTICLE

Ross McKitrick

THERE are, sometimes, cheap shots taken at the United Nations Intergovernmental Panel on Climate Change (IPCC); there are commentators who, without having read any of its reports, simply attack it on political grounds. Unthinking hostility towards the IPCC is as harmful as unthinking adulation. The IPCC addresses itself to serious but very contentious scientific matters and, in principle, aims to do so in a conscientious and professional way. But it is not guaranteed to be correct, nor do its procedures guarantee that it is unbiased and properly representative of the full spectrum of professional views. David Henderson's presentation in this conference volume politely but firmly calls upon governments to consider what would be the consequences if the IPCC reports are biased or flawed. We ought to approach such a question without a presumption of guilt, but we should not be afraid to ask the question. As the saying goes, *trust but verify*.

Questions about the competence and objectivity of the IPCC matter, not only because of the importance of the climate change issue, but because the IPCC occupies an institutional niche as the designated monopoly supplier of scientific advice for countries under the UN Framework Convention on Climate Change (UNFCCC). Wording in IPCC reports has major implications for global energy policy, and if conclusions were to be issued by the IPCC that trigger Article Two of the UNFCCC, then its conclusions would effectively carry the force of international law. Therefore, I agree with David Henderson that critical scrutiny of the IPCC is much needed, with particular reference to oversight mechanisms. Considering the IPCC's international role, it is remarkable that the only oversight is managed by the IPCC itself. A question to motivate our line of inquiry is: Suppose a journalist or researcher uncovered incontrovertible evidence that the IPCC had deliberately falsified data during the final preparation of one of its *Assessment Reports*. What is the phone number of the agency that could be called upon to investigate and, if necessary, prosecute? And what court would have jurisdiction?

In my comments I will explain why I believe the core group that influence the IPCC's reports and conclusions is biased toward the view that greenhouse gases are the cause of major, deleterious global warming, and why I think this bias leads that group to censor or even misrepresent opposing evidence. I will draw on my interactions with Working Group I over its use of my research and that of others with which I am familiar. But I think David's proposed remedies are not adequate. Any solution needs to create strong incentives for the IPCC to fix itself. After presenting evidence of IPCC bias, I will use some economic reasoning to propose a fix for the IPCC that would, I hope, not simply replace its biases with different biases but would make it more truly objective and balanced.

Examples of Bias in the IPCC Process

There are many layers to the IPCC. At its core are a few dozen individuals who serve as IPCC Bureau members or Lead Authors and who exert considerable control over the contents and conclusions of the report (chiefly the Working Group I contribution to the *Fourth Assessment Report*, or AR4, with which I am concerned here). Around them, like concentric circles, are larger and larger groups of contributing authors, reviewers and government reviewers, until the widest circle encompasses many hundreds, perhaps thousands, of people. I served as an expert reviewer, by invitation rather than self-nomination, for what that's worth. I found—admittedly to my surprise—that many sections of the IPCC report are truly well done. They reflect credible expert contributions from knowledgeable sources. However, these sections don't grind axes or tell scary stories: they report on advances in scientific understanding, while making clear how complex and difficult the study of climate remains. This is not the stuff on which the Summaries and subsequent headlines get built.

In key places the tone of the report changes. It becomes brittle, argumentative, and slanted. Here the heavy hand of the core writing team takes control, on topics that dominate the overall conclusions and the Summary for Policy Makers (SPM). Since contributors and reviewers are never asked to vote on whether they agree with the SPM, any suggestion that the core writing team speaks for the thousands of people in the wide concentric circles is misleading.

I have published on some of these key topics, and I followed the IPCC's drafts on them closely. In every case, partisans on the alarmist side of current controversies were asked to summarize the debates, an obvious conflict of interest, resulting in tendentious and incomplete discussions. I will discuss a few examples, and then I will propose how to make the policy process robust toward the possibility that the IPCC were wrong and even give the IPCC an incentive to start getting things right.

Example One: Surface Temperature Record

It is intuitively obvious (and empirically well-established) that the growth of cities, and other transformations of the earth's landscape, can cause a rise in local temperatures. Climatic data is supposed to be adjusted so that it only measures a "pure" air temperature signal, in effect showing what the temperature in a region would have been if there had never been any human settlement in the area. Treating published climate data as if they accurately reveal this hidden information requires some heroic assumptions. Since climate models do not predict a spatial pattern of warming that matches the spatial pattern of industrial development, there is a simple test of how successful the data adjustment models are. If climate data are uncontaminated, the spatial pattern of warming trends over land should be uncorrelated with the spatial pattern of industrial development and other indicators of measurement quality. The IPCC and other authors have long asserted this independence. Indeed their conclusions about warming and the detection of a greenhouse gas influence presuppose it. But until a few years ago nobody had ever tested it.

There were three studies published after the Third Assessment Report, one coauthored by me (McKitrick and Michaels 2004) and two by the team of de Laat and Maurellis (2004, 2005) that tested the hypothesis. We worked independently and published our papers, not knowing about each others' work. We used independent data sets and approached the topic with different statistical methods. We all concluded, with overwhelming statistical significance, that the IPCC's climate data are contaminated with surface effects from industrialization and data quality problems, which together add up to a large warming bias on the order of half the observed warming over land since 1980.

In their first two drafts, the IPCC simply ignored these papers. They referred instead to an 18-year-old paper by IPCC Lead Author Phil Jones that had failed to find an urbanization bias in a few regions of Asia and Australia; a similar study for the United States by Peterson; and two recent papers by IPCC Lead Author David Parker which compared the strength of urban heat islands on windy and calm nights, finding little difference. On this basis they

85

asserted quite categorically that their climate data is uncontaminated by a warming bias due to land surface changes or measurement problems.

This question goes to the very heart of the IPCC's position. If the locations of maximum warming coincide with the locations of socioeconomic development, and if this is not predicted by climate models as a consequence of greenhouse gases, it would imply that the fundamental data set used by the IPCC is contaminated and would put into question a host of their conclusions about the extent of observed warming and its attribution to greenhouse gases. The topic deserved an extensive and prominent treatment, but instead it was hurriedly dismissed.

In my expert review comments I drew attention to my paper and those by de Laat and Maurellis and rebutted the arguments based on the Jones paper. In the second draft, nothing changed. The IPCC continued to ignore the topic, so I reiterated all my criticisms in the second round, leading to the following response by the Lead Authors (Chapter 3, Second Draft Review Comments, line 3-453):

> Rejected. The locations of socioeconomic development happen to have coincided with maximum warming, not for the reason given by McKitrick and Mihaels [sic] (2004) but because of the strengthening of the Arctic Oscillation and the greater sensitivity of land than ocean to greenhouse forcing owing to the smaller thermal capacity of land. Parker (2005) demonstrates lack of urban influence.

Elsewhere the Lead Authors dismissed my paper by saying it is "full of errors," without providing any evidence or details.

The appeal to the Arctic Oscillation to explain away the warming pattern in land-based surface temperatures across both hemispheres is ridiculous. Where the IPCC discusses causes of observed climate change, the Arctic Oscillation is not even invoked to explain *Arctic* warming, let alone warming in South America or Africa. In Chapter 9, which discusses attribution of climate change to anthropogenic factors, the IPCC report makes only one mention of the Arctic Oscillation (p. 693), where it is noted that excluding it from an analysis has no effect on the results. In discussions of factors behind Arctic temperature changes (pp. 694, 714, 716), Arctic Oscillation is not mentioned at all.

It is highly ironic that, confronted with published, peer-reviewed evidence of an anthropogenic, but non-greenhouse effect on temperature trends, the

IPCC denied it by appealing to natural causes. In the end, the report seemed to conclude that its authors were obliged to discuss the topic, but the text they published turned out to be misleading. In the published version of the AR4 (Chapter 3, page 244), a paragraph was inserted regarding the three papers in question (though only the latter de Laat and Maurellis paper was cited):

> McKitrick and Michaels (2004) and de Laat and Maurellis (2006) attempted to demonstrate that geographical patterns of warming trends over land are strongly correlated with geographical patterns of industrial and socioeconomic development, implying that urbanization and related land surface changes have caused much of the observed warming. However, the locations of greatest socioeconomic development also are those that have been most warmed by atmospheric circulation changes (Sections 3.2.2.7 and 3.6.4), which exhibit large-scale coherence. Hence, the correlation of warming with industrial and socioeconomic development ceases to be statistically significant. In addition, observed warming has been, and transient greenhouse-induced warming is expected to be, greater over land than over the oceans (Chapter 10), owing to the smaller thermal capacity of the land.

Note the slanted language: "*attempted* to demonstrate" instead of "*showed*." This is an example of the subtle bias created when partisans get to write the report summarizing disputes of which they are a part.

The paragraph effectively admits the existence of evidence that the spatial pattern of warming coincides with the spatial pattern of industrial development but attributes it to natural atmospheric circulation changes, referring the reader to two subsequent sections. Neither section shows any such thing—the overlap between warming patterns and socioeconomic development is simply not mentioned in those sections. The claim that our results become statistically insignificant when this effect is controlled for is a pure fabrication. Our papers show no such thing, and neither do the cited sections, nor does the IPCC have any statistical evidence to back up its assertion, nor can it point to any peer-reviewed publication for support. The IPCC's claim is simply made up.

Thus, on one of the most important topics raised in the report—published evidence of significant contamination of the primary data set used for measuring global surface warming over land and testing for an influence of greenhouse warming—the IPCC authors initially dismissed published, peer-reviewed evidence by appeal to an irrational, ad hoc speculation about the Arctic Oscillation and then fabricated a claim that the evidence against their position was statistically insignificant. I consider this to be a serious failing on the part of the IPCC.

Example Two: The Paleoclimate Record

In the *First Assessment Report* (1990), the IPCC declined to conclude that 20th century warming could be attributed to humans, citing an apparent warming interval in the medieval era as a significant obstacle to drawing such a conclusion. They published a schematic showing that there was a significant warm period from roughly 980-1370. (See Figure 1.)

In the IPCC's *Third Assessment Report* (2001), this history was swept away in light of the now-famous "hockey stick" graph of Michael Mann, which appeared prominently in the Summary for Policy Makers and at least four other places in the report. The graph appeared to show that there was no medieval warm period and that the 20th century climatic change was beyond all bounds of natural variability. The reported medieval temperatures were, below normal or declining, roughly 1370-1870. (See Figure 2.)

The remarkable visual shape is not characteristic of Mann's proxy library. The simple average of the proxies in his data set looks nothing like a hockey stick and does not even slope upward in the 20th century. The hockey stick shape emerged solely as a consequence of the way the data were averaged. (See Figure 3.)

Figure 1

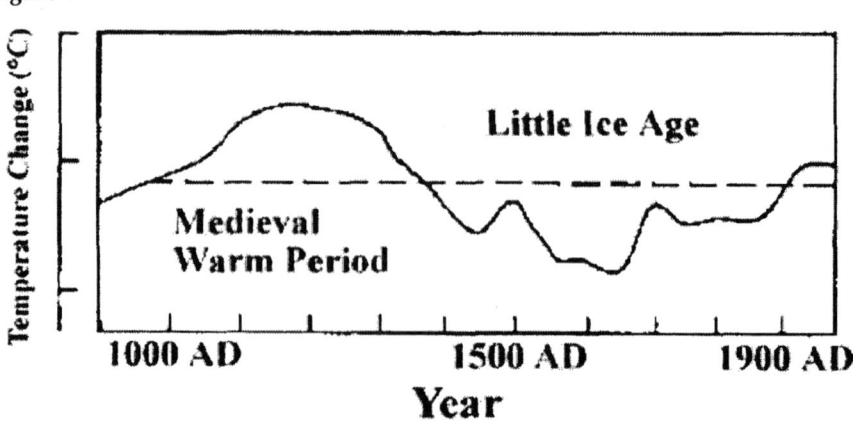

Source: *First Assessment Report*, IPCC, 1990.

Figure 2

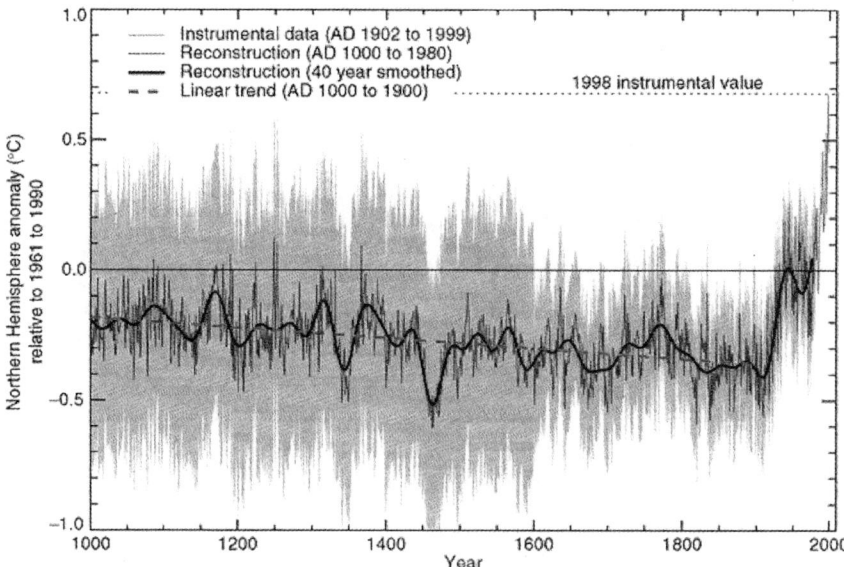

Source: *Third Assessment Report*, IPCC (2001) http://www.grida.no/climate/ipcc_tar/slides/05.16.htm.

Figure 3

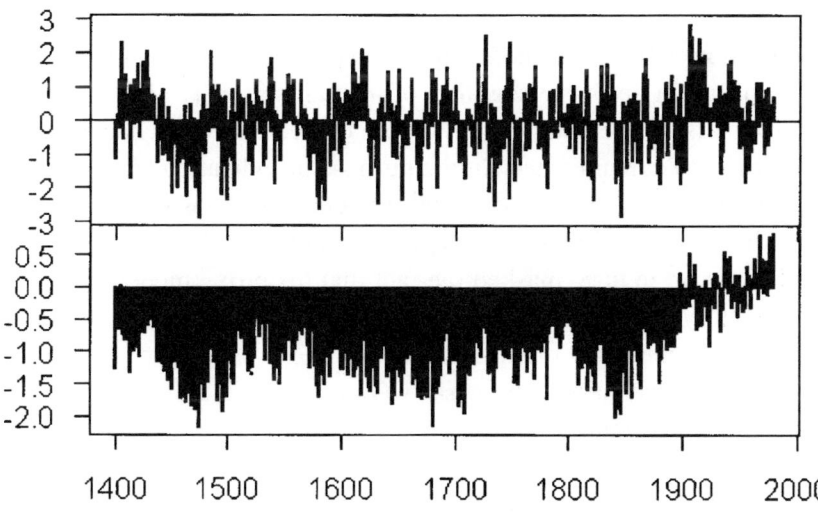

Top— Simple average of 415 proxy series in Mann's data set.
Bottom hockey stick reconstruction.

Source: McIntyre and McKitrick submission to the US National Academy of Sciences expert panel on paleoclimate reconstructions, http://www.uoguelph.ca/~rmckitri/research/NAS.M&M.pdf

Figure 4

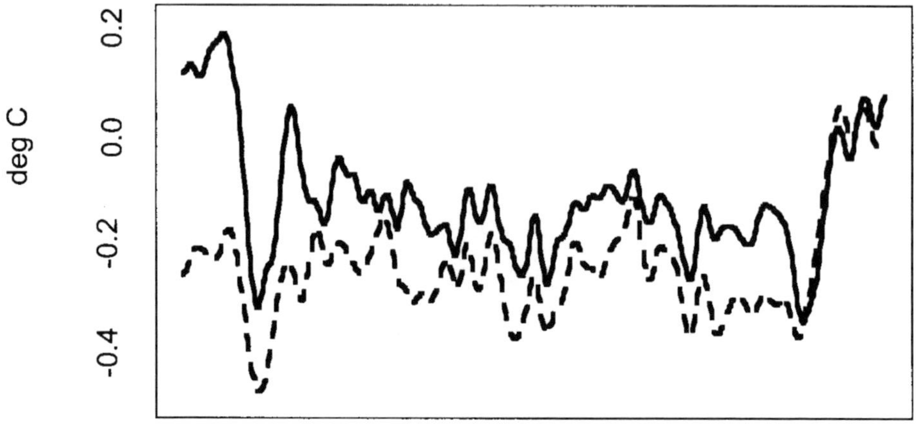

The hockey stick (dashed) with the bristlecone influence removed (solid).
Source: Stephen McIntyre, personal communication.

As is now well-known, and cannot be reviewed here in detail,[1] Stephen McIntyre and I published a series of articles that demonstrated calculation errors in the Mann analysis that undermined his conclusions.

- We showed that he applied an incorrect principal component algorithm, resulting in an over-weighting of a small set of tree ring indicators from bristlecone pines in the Western United States that had long been viewed as contaminated for the purpose of indicating historical temperature due to the predominant influence of atmospheric chemistry rather than temperature on its growth rates.
- This error, in turn, masked the fact that his proxy model had no explanatory power for past temperature (that is, was no more informative about the past than random numbers).
- The error also masked the sensitivity of his results to the use of bristlecones. (See Figure 4.)

In light of the controversies raised by our findings, the U.S. Congress asked the National Research Council (NRC) and the Chairman of the National Academy of Sciences Committee on Theoretical and Applied Statistics to set up separate panels to review the matter.[2] Both panels accepted McIntyre's and my arguments and produced independent numerical replications to confirm our mathematical results. Both panels criticized the hockey stick, the second panel (the Wegman panel) doing so in particularly severe terms. The NRC panel went into some wide-ranging examination of tree ring-based paleoclimatology and concluded, among other things, that the pre-1600 era is "murky" for the purpose of comparing to the present day. They specifically cautioned against relying on strip-bark formations in bristlecone pines, a type of proxy that underpins all paleoclimate climatic reconstructions used by the IPCC since 2001 to argue against a relatively warm medieval era. Without strip-bark samples, these reconstructions would all show the present climate to be unexceptional compared to the medieval warm period. They also accepted our findings that the claims of significant correlations between proxy series and temperature data were unfounded.

The sequence of IPCC drafts, reviewer responses, and the final wording has been reviewed in a remarkable new paper by David Holland, to be published in 2007 *Energy and Environment*, which I strongly recommend for anyone who wants some insight into the real workings of the IPCC.[3] What Holland recounts, based on the publicly available record of IPCC drafts and reviews, demands a verdict of either profound stupidity or deliberate misrepresentation on the part of the IPCC Chapter 6 Lead Authors.

Despite McIntyre and McKitrick having published five journal articles on the hockey stick controversy by the time the IPCC report was being drafted, the IPCC initially ignored all but our first paper. They falsely claimed that we had offered up a novel climate reconstruction that had failed model validation tests and that we had been unable to replicate Mann's work because we omitted a key part of his data set. They also claimed that our results were rebutted in an unpublished paper by Wahl and Ammann, who had (they said) successfully replicated Mann's results. As we pointed out in our replies, none of this was true. We had repeatedly denied that we were presenting a new reconstruction; instead we were attempting to replicate Mann's reconstruction based on his stated methods and data. We showed that it was not possible to get his results using his stated data and methods. The IPCC failed to mention that we had proved to *Nature*'s satisfaction that the original disclosure of data and methods was, indeed, inaccurate, and a *Corrigendum* had been ordered.[4] Based on the amended disclosure of data and methods, the results of Wahl and Ammann were identical to *ours*, not to Mann's, and like us, Wahl and Ammann had found that Mann's claims of finding statistical significance could not be replicated. But the draft version of the Wahl and Ammann paper submitted to the IPCC omitted the latter findings, which were included in the version they had submitted to a journal for publication. And their paper had not been published, which should have ruled out its usage by the IPCC in any case.

We (and others) submitted detailed critical replies rebutting the IPCC's summary. In their notes to the Review Editors the chapter authors appeared to concede the criticisms and said the text would be edited, but the Second Draft was almost identical to the first. The IPCC repeated its mischaracterization of our work and continued to claim Wahl and Ammann's (still unpublished) paper refuted our claims. They also added a grudging acknowledgment that our critique of the flawed principal component analysis "may have some foundation," but they dismissed it anyway, despite both the NRC and Wegman panels having upheld our analysis.

We, and others, filed further objections to this section. The U.S. Government Reviewer noted that the Wahl and Ammann paper had missed the IPCC's final, extended deadline for inclusion (it remains unpublished to this day) and ordered the removal of all references to it. The Lead Authors ignored this request, flouting the IPCC rules in the process, and the Review Editors apparently rubber-stamped their decision.

The final, published text of the IPCC report thoroughly misrepresents the hockey stick debate, ignores published evidence against Mann's original results that had been upheld by two independent expert panels, relies on unpublished claims in the Wahl and Ammann paper while ignoring their replication of our results, etc. This whole section of the AR4 is indefensible and stands as a lasting testament to the bias of its authors and the willingness of the IPCC process to indulge such biases.

Long Term Persistence

Detection of a trend in a data set, such as temperature records, requires more than simply fitting a line through it. The researcher has to decide if the trend is "large" or not; that is, whether it is statistically significant, or outside the bounds of mere noise. This can be quite difficult to do since there are many properties of time-series data that can cause the standard trend estimation formulas to give answers that are biased toward calling a trend significant when in reality it is not. For instance, most students of introductory time-series analysis learn about the problem of *autocorrelation*, in which processes with an intrinsic "momentum" are slow to respond to random fluctuations. Autocorrelation can cause trend estimation methods to overstate significance.

A form of autocorrelation often observed in geophysical data is called Long Term Persistence (LTP). LTP models arose in hydrological studies of long-term Nile River depth records. Researchers noted that hydrological events (droughts, floods) tended to cluster together over time, causing long-term excursions in the data that appeared as pseudo-trends over short intervals. Standard time-series analysis based on simple autoregressive models do not adequately capture this effect, so LTP models for geophysical data series have been developed. A substantial body of empirical work has been published in recent years showing that many basic climatic processes exhibit LTP, and classical statistical methods will lead to over-estimation of the significance of trends. Cohn and Lins (2005)[5] developed a test for the significance of trends in geophysical data that is robust toward the presence of LTP. They showed that what appears to be a highly significant upward trend in a common "global temperature" series under the autoregressive assumption that falls to insignificance when the test allows for LTP. They conclude:

> [With respect to] temperature data, there is overwhelming evidence that the planet has warmed during the past century. But could this warming be due to natural dynamics? Given what we know about the complexity, long-term persistence, and non-linearity of the climate system, it seems the answer might be yes. . . . Natural climatic excursions may be much larger than we imagine.

While I have not (yet) published on LTP, I have served as referee for a few climatology journals on the topic. In my capacity as an IPCC expert reviewer I objected to the IPCC's use of simplistic, obsolete methods to evaluate the significance of trends in temperature data sets. In response to the first draft of the report, another reviewer and I drew attention to the literature on LTP phenomena, asked that it be properly referenced, and requested that the temperature trend significance calculations be redone using correct, up-to-date methods that followed the peer-reviewed literature. The chapter authors were antagonistic to this suggestion and refused to change their methods, but the second draft of the IPCC report did, at least, introduce a short discussion of the LTP issue as follows (Second Draft, pages 3-9):

Determining the statistical significance of a trend line in geophysical data is difficult, and many oversimplified techniques will tend to overstate the significance. Zheng and Basher (1999), Cohn and Lins (2005) and others have used time-series methods to show that failure to properly treat the pervasive forms of long-term persistence and autocorrelation in trend residuals can make erroneous detection of trends a typical outcome in climatic data analysis.

A similar comment was inserted in the chapter appendix, though it included a disputatious and unsupported assertion that LTP models lack physical realism, to which I presented a counterargument in my second draft review comments.

The paragraph above makes an important point that has direct bearing on the overall conclusions of the IPCC report. Then without explanation, that paragraph was deleted from the published edition. The entry in the Appendix was made even more disputatious, even though no supporting citations were provided for the dismissal of Cohn's and Lins's results.

Level of Scientific Understanding

There are a few other areas where I have noted a pattern of bias in the IPCC but which fall outside my own research areas. One is the evolution of "Scientific Understanding" ratings among drafts. There is a diagram in the Working Group I Summary for Policy Makers (SPM) that sums up the estimated contributions of different planetary variables to the so-called "radiative forcing" of climate. Accompanying each one is an assessment of the "Level of Scientific Understanding" or LOSU. All the listed entries are Low, Medium, or High. In the review comments on the Second Order Draft,[6] comment number 2-1273 read:

It is notable (surprising?) that the level of scientific understanding for pre-satellite-era solar forcing which is based on proxies and models has jumped from "Very Low" in the [*Third Assessment Report*], to "Medium" in the AR4 figure. This should either be explained and highlighted here, or corrected including in this Figure which appears 3 times. In addition, this contradicts Chapter 2, page 6, lines 27-28!

The author response was:

Changed to "low." Accepted.

94

The difference between "Very Low" and "Medium" for a category as important as solar influence on climate implies quite a substantial difference in scientific understanding, yet is decided here by what amounts to haggling between a reviewer and an author. In other words, Lead Authors do not even claim enough scientific understanding to decide what the level of scientific understanding is. As will be pointed out below, the AR4 surveys historical solar proxies and finds a wide range of results with widely-varying implications for understanding the solar influence on climate. Had the reviewer not drawn attention to this item, the LOSU would have been listed as Medium; because of one objection it was scaled down to "Low," suggesting that the authors had no basis for scaling it up so far in the first place.

It is also interesting to look at the way the LOSU ratings were inflated between the second and final LOSU drafts—see Table 1. In the first draft, 6 out of 15 climate forcing categories were rated as Very Low scientific understanding. In response to reviewer comments, the second draft scaled down its certainty ratings so that 7½ out of 15 were Very Low (contrails includes two sub-categories, one Low and one Very Low). In other words, half the categories of major climatic forcings were subject to the lowest possible rating for scientific certainty. I did not find any review comments on the second draft saying this overstated the uncertainty, yet in the final,

Table 1: Evolution of "Level of Scientific Understanding" ratings in IPCC *Fourth Assessment Report* Table 2.11, and in the Summary for Policymakers Figure 2. *"-"* denotes *not shown*.

Forcing Category	Level of Scientific Understanding (LOSU)			
	1st Draft	2nd Draft	Final Draft	SPM
Greenhouse gases	H	H	H	H
Stratospheric and Tropospheric ozone	M	M	M	M
Stratospheric water vapor from methane	L	L	L	L
Stratospheric water vapor from other	V. L	V. L	V. L	-
Tropospheric water vapor from irrigation	V. L	V. L	V. L	-
Aerosol scattering and absorbing	L-M	L	L-M	M-L
Cloud albedo effect	L	V. L	L	L
Cloud lifetime effect	V. L	V. L	-	-
Cloud semi-direct effect	V. L	V. L	-	-
Contrails and aviation cirrus	M	L-V.L	L	L
Solar	M	L	L	L
Cosmic Rays	V. L	V. L	V. L	-
Surface Albedo	L	L	M-L	M-L
Non-Albedo Surface	V. L	V. L	V. L	-
Volcanic	M	L	L	L
Proportion Listed as Very Low	6 / 15	7½ / 15	4 / 15	0 / 8

published report only 4 of 15 Very Low ratings are shown (with two categories deleted). Also, in the Summary for Policy Makers Figure SPM-2, none of the forcings in the Very Low categories appears, creating the impression of greater certainty than was indicated in Table 2.11 at the close of scientific review.

Solar Reconstructions

The IPCC acknowledges that solar activity is high, and possibly exceptionally high, compared to the last 8,000 years. The two most prominent proxy-based reconstructions (from teams led by Solanki and Muescheler, respectively), differ on whether an interval in the 1700s included a spike comparable to today's but both agree that today's solar output is very high compared to most of the current interglacial era.

There have been many reconstructions of total solar activity based on sunspot counts, which began in the early 1600s. Until recently, most reconstructions showed a strong upward trend in total solar irradiance since the 17th century, with low-frequency trends that track 19th and 20th century average temperatures reasonably well. Very recently, however, a different-looking reconstruction by Wang et al.[7] suggested that climatic forcing due to total irradiance had risen very little since the 1700s, implying an increase in solar output as little as one-tenth the size reported by other solar reconstructions. This would imply that solar changes could have little to do with climate change since 1600.

Rather than treating the Wang et al. results as one of a range of recent findings, it was presented as the sole and sufficient basis for a conclusive dismissal of the larger solar forcing estimates used in earlier IPCC reports, a dismissal that carried over into the Summary for Policy Makers, where it is concluded that solar influence on climate is much smaller than has been proposed earlier. Here the bias amounts to cherry picking. Readers are told that one new result, standing at odds with a host of earlier studies, is the definitive word on the matter. Had the one new result found evidence for much stronger solar forcing than earlier thought, it is likely that it would hardly have been mentioned, or that it would have been presented in a disputatious and grudging side comment.

Tropical Troposphere

Global climate models implement a hypothesis in which strong

Figure 5

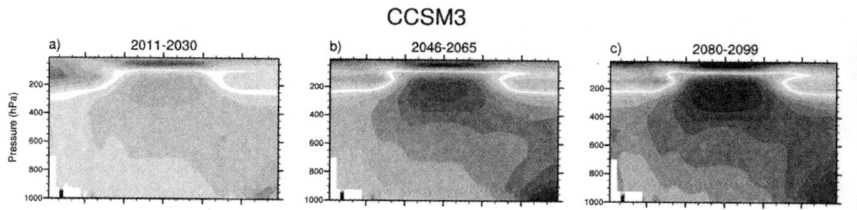

infrared absorption by CO_2 raises the effective emissions altitude and forces a warming response centered in the tropical troposphere. IPCC AR4 Figures 9.1 and 10.7 confirm the importance of this mechanism by showing that repeated model runs, for historical and future intervals, are dominated by tropical tropospheric warming.[8] The CCSM3 diagrams depicting predicted tropospheric temperature trends in selected 20-year intervals show a marked increase of warming as one goes from 2011-2030 to 2046-2065 and then to 2080-2099. (See Figure 5.)

In each of the three panels, the horizontal axis shows latitude (left to right = South Pole to North Pole) and the vertical axis shows altitude, measured in atmospheric pressure, corresponding to about 25 km total height. The shading denotes the intensity of the warming trend (darker shading denotes greater intensity). The three panels refer to three time intervals. In Figure 9.1 the IPCC generates a profile very similar to its prediction of what ought to be observed already in 20th century data.

The big dark boil in the top of the middle diagram is the tropical troposphere. Although a small region in the diagram box, over the globe it takes up one-half of the lower atmosphere. In all model runs, this is where the warming starts and runs strongest. Figure 9.1 indicates that this pattern is uniquely associated with greenhouse gas (GHG) effects. Amplified warming at the North Pole is associated with GHG emissions but also is associated with solar changes. The amplified warming over the tropics is associated only with greenhouse gases.

A similar hindcast is in the U.S. Climate Change Science Program (CCSP) Report, page 25.[9] That report also shows that the tropical troposphere already should be warming.

It is remarkable, therefore, that the IPCC did not plot the available data on the tropical troposphere in the same format as the model outputs above in order to allow visual comparison and evaluation. The data they do show

Figure 6

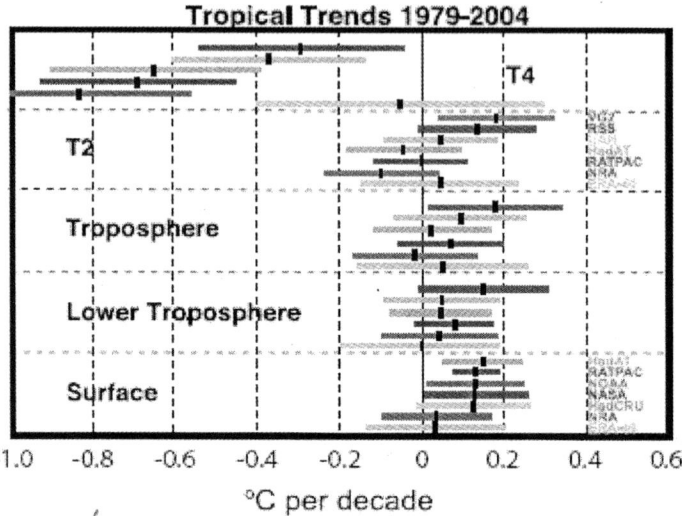

Copied from IPCC AR4, Working Group I, p. 269.

(AR4 Figure 3.18) is in a different and visually unobtrusive format. But its meaning is clear enough: across the six data sets there is essentially no evidence of significant warming in the tropical troposphere. (See Figure 6.)

The meaning of this data set is drawn out as follows on page 11 of the US CCSP report on reconciling surface-tropospheric data (emphasis added):

> For global averages (Fig. 3), models and observations generally show overlapping rectangles. A potentially serious inconsistency, however, has been identified in the tropics. Figure 4G shows that the lower troposphere warms more rapidly than the surface in almost all model simulations, while, in the majority of observed data sets, the surface has warmed more rapidly than the lower troposphere. In fact, the nature of this discrepancy is not fully captured in Fig. 4G as *the models that show best agreement with the observations are those that have the lowest (and probably unrealistic) amounts of warming.*

Take out the "probably unrealistic" gloss and this is an admission of something pretty significant. The IPCC makes almost no mention of the issue. Had there been evidence of rapid warming in the tropical troposphere, the data would have been presented in a prominent format, accompanied by extensive discussion. The decision to downplay this topic is another indicator of the biased editorial hand guiding the final report.

How to Fix the IPCC

I could go on with other examples, but I hope by now to have justified for the reader my belief that the core writing team of the IPCC *Report* shares a single point of view, that its members are alert and predisposed toward evidence that confirms that point of view, and that they are unreceptive or openly hostile to evidence that contradicts it. Whether the reader agrees or not, one should take it as given, for the moment, that there is a problem of bias.

Now I would like to turn to the question of how to fix the situation. David Henderson outlined some suggestions that seek to increase oversight and broaden the milieu. I agree that increased oversight may help. Together with Bruce McCullough, an econometrician at Drexel University, I have just finished a report for the C.D. Howe Institute[10] documenting the need for due diligence when empirical research is used in policy formation. While we mainly focus on the problem of pervasive non-reproducibility of economics research, the remedies we propose would also cover research from environmental scientists and others. Creating a mechanism for checking whether published research is reproducible (i.e., whether the data and methods have been accurately and fully disclosed) would catch a lot of climate science in its net.

But as for bringing in other branches of government to try and balance the process, my concern is that the milieu would simply swallow up new entrants, alive and whole. The incentives just are not present for other divisions of the government to question the IPCC in any serious way. If a team of economists at, say, the U.S. Department of Commerce were asked to weigh in on the climate change issue, it would quickly become apparent to them that their troubles would be minimized by conceding the scientific grounds entirely and confining their comments to some narrow economic issues. If they were to weigh in on, say, statistical methods in greenhouse gas "signal detection" methods, even if they had as many or more qualifications to do so as their colleagues in an environmental department or bureau, they would know (or soon learn) that any challenge would be met with an extremely demanding counter-attack. They would have to devote all their energies for several months to answering a barrage of responses, usually lobbed with considerable public hostility and vituperation, and even if they succeeded in answering them, the targets of their criticism would retain the option of carrying on as if nothing had changed, safe in the assumption that the IPCC will not demote their views as long as they continue to get them into print in any form, and, in a pinch, even if they cannot get them into print.

The alternative of keeping one's head down and sticking to one's own knitting would save a lot of bother and would be an irresistibly attractive alternative. Because these incentives would likely prevent any serious oversight from adjacent government agencies, I do not believe any solution will emerge from within the government or international government agencies.

Ultimately, reform cannot be imposed from the outside. The core IPCC leadership must *want* to be a neutral, accurate, and honest information source. If they are ultimately interested in promoting a policy agenda, then they will always find a way around any attempt to force them to be balanced. So I would prefer a policy that would create incentives for accuracy in the IPCC.

My proposed solution will, at first glance, appear to have nothing at all to do with reforming the IPCC. It begins with a proposal for a carbon tax. Suppose a government—anywhere in the world, but preferably one with a large industrial economy like the United States—imposes a carbon tax *whose value is tied to the mean temperature of the tropical troposphere*, averaged over two or more of the data sources shown in Figure 3.18 of the AR4, Working Group I *Report.*

The IPCC has predicted that this region of the atmosphere is supposed to lead the global warming process, *if it is caused by greenhouse gases.* Other regions of the planet (especially at the Northern Hemisphere surface) might warm due to circulation changes, urbanization, or other factors. But IPCC climate models suggest that only one thing will cause a sustained, pronounced warming in the tropical troposphere, greenhouse gases.

So, suppose the U.S. government (or any other government) implements a low carbon tax, with the revenue recycled locally, calibrated to that temperature measure. I would call it the T3 tax, for "tropical tropospheric temperature." If the mean tropospheric temperature starts going up, the T3 tax would go up, forcing emissions down. If the tropical troposphere does not warm up, the tax won't go up, nor should it.

I have spelled out the research behind this proposal in some essays available at ross.mckitrick.googlepages.com. Consider this formula:

$$T3 = 20 \times \tfrac{1}{12} \sum_{i=0}^{11} \tfrac{1}{2}\big(SC(t-i) + RSS(t-i)\big) \qquad (1)$$

where $SC(t)$ is the Spencer-Christy monthly mean tropical tropospheric temperature anomaly and $RSS(t)$ is the same from the Remote Sensing Systems (RSS) laboratory. Equation (1) says that the tax rate should be set equal to 20 times the twelve-month moving average of the mean of the RSS and University of Alabama-Huntsville (UAH) estimates of the mean tropical tropospheric temperature anomaly. By using a one-year trailing average, the movements would be smoothed out, limiting spikes or drops due to, for example, volcanic activity or strong El Nino events.

Based on current data (as of August 2007), the T3 tax would be about US $4.70 per ton of carbon, and it should be increasing by between $4 and $24 per decade, according to the range of IPCC projections. Its post-1980 historical trend is less than that, at around $3 per decade. It would exhibit far less volatility than the European carbon market price, which has swung between $0 and $30 (US) per year since inception. (See Figure 7.)

Politically, an advantage would be that alarmists and skeptics alike should expect to get their preferred outcome. Alarmists will expect a rapidly accelerating carbon tax rate, while skeptics will expect to see a carbon price that stays low for a while, then possibly drops when we enter solar cycle 25 around 2020, which is expected to be marked by diminished solar output.

Economically, such a carbon emissions tax has the advantage of being

Figure 7

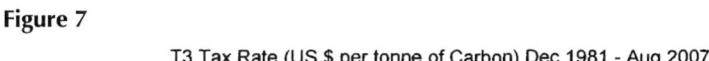

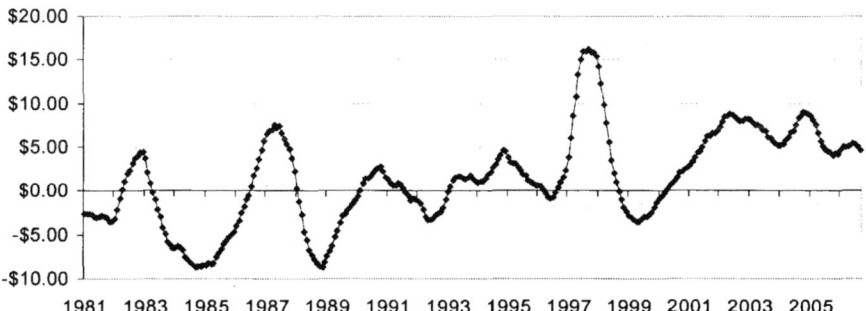

Formula (1), using the Spencer-Christy (University of Alabama) data and the Mears-Wentz (Remote Sensing Systems) data to compute the carbon tax.

a clear price mechanism. Other presentations at this conference by Gilbert Metcalf, Robert Mendelsohn, and Peter Wilcoxen, will explain in more detail why choosing a carbon price is a particularly efficient way to achieve emissions reductions. This formula simply pins the rate to an objective measure of the effects of greenhouse gases. Whether the rate goes up or not, we will end up with the right outcome, without having to guess in advance what the right policy is.

As for the purpose of reforming the IPCC, note that the T3 tax would require firms to form long-term expectations about future climate change to guide today's decision-making. Someone building a pulp mill or a power plant would have to get the best information available about climate trends for the next ten or twenty years in order to project the carbon price they would face. They do not want to know what today's value of the T3 tax is; they will want to know what it will be ten, twenty, or thirty years down the road. This will create a market for *accurate* and *objective* climate analysis and forecasts. Firms with large investments on the line will have an incentive to get the tropospheric forecasts right. This will force them to peel away the layers of bias and drill down to objective science.

My conjecture is that a new consensus forecast would emerge that shows relatively little tropical tropospheric warming for at least the coming few decades. In other words, I predict that the *market* would say that the IPCC is wrong. Its judgment would be based on the decisions of neutral third parties whose job is to forecast a tax rate that would guide hundreds of millions of dollars of private sector investment. Whoever holds that job will not have an inch of room to indulge his or her prior biases on global warming. Forecasters will have to become skeptical, dig into the data, ask hard questions of modelers, and persuade their clients that they are giving trustworthy forecasts. Whether the forecaster is a true believer in global warming or not, he or she will know that clients need the right answer, regardless of whether it is politically correct or not, and the market will be looking for someone who can establish a track record of valid climate forecasting, regardless of what theory guides the modeling.

Over time, it is possible that the IPCC's analysis and forecasts would be vindicated. But I doubt it. I think that over time the IPCC would be seen as an outlier with a track record of exaggeration. At that point, the IPCC would have to reform itself or risk oblivion. Establishing a policy framework which creates competition and rewards objectivity and accuracy will do more than anything to fix the problem of bias in an intellectual monopoly like the IPCC.

Endnotes

[1] See http://ross.mckitrick.googlepages.com/#hockeystick for a collection of explanatory papers.

[2] Copies of the reports are available at http://www.uoguelph.ca/~rmckitri/research/trc.html.

[3] It is available online at http://tinyurl.com/2szwh8.

[4] It was published in *Nature*, July 1, 2004, p. 105.

[5] Cohn, T.A. and H.F. Lins, "Nature's style: Naturally trendy," *Geoph. Res. Lett.*, vol. 32 (2005), L32402, doi:10.1029/2005GL024476.

[6] http://ipcc-wg1.ucar.edu/wg1/Comments/wg1-commentFrameset.html.

[7] Wang, Y.M., J.L. Lean, and N.R. Sheeley, "Modeling the sun's magnetic field and irradiance since 1713," *Astrophysical Journal*, vol. 625 (2005), pp. 522-538.

[8] For 10.7 see all 12 global climate model runs at http://ipcc-wg1.ucar.edu/wg1/Report/suppl/Ch10/Ch10_indiv-maps.html.

[9] http://www.climatescience.gov/Library/sap/sap1-1/finalreport/default.htm.

[10] McCullough, Bruce D. and Ross R. McKitrick, "The Case for Due Diligence when Empirical Research is Used in Policy Formation," C.D. Howe Institute *Commentary* (Forthcoming, 2008).

GLOBAL WARMING AND THE UNITED NATIONS

Claudia Rosett

FIRST, I thank you in advance for your fortitude in being willing to sit through a talk with the United Nations in the title. I want to start by showing you the chart in Figure 1, which gives a sample of the complexity of the relevant part of the organization.

The dates indicated are the years the secretariats were set up, but please remember that sometimes the relevant treaties were signed earlier.

When Walker Todd and I first discussed this presentation, the title we agreed on involved the United Nations, the Intergovernmental Panel on Climate Change (IPCC), and "money trails," and I intend to add that element to the discussion, offering a different prism for viewing the global warming issue.

I am not a scientist. But having listened to the fascinating presentations this morning, it does strike me that we have heard that the study of climate

Figure 1. United Nations Entities Dealing with Climate Change Issues.

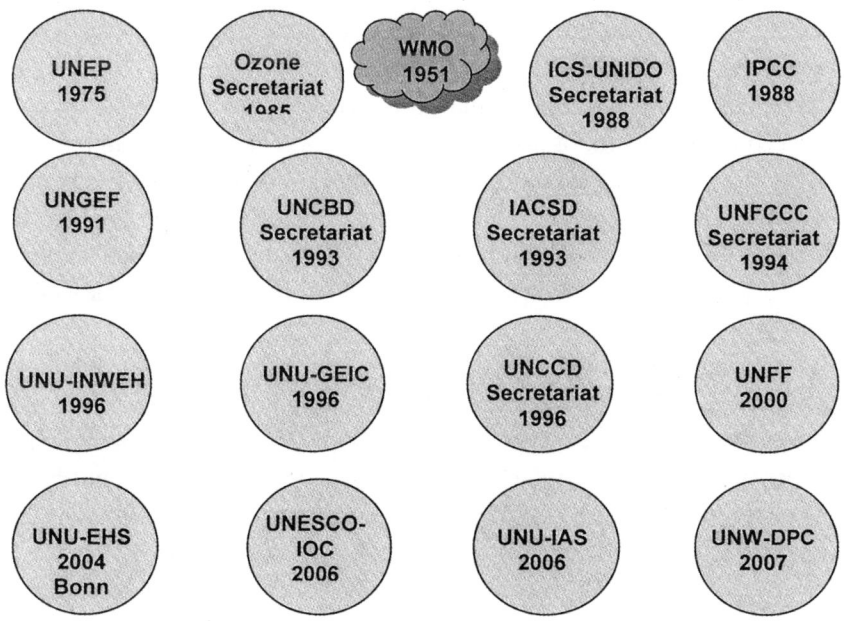

change entails many more variables than we know how to solve. When faced with that circumstance in normal human experience, the answer often is to try to arrange options so that, when something unexpected happens, one has the flexibility to deal with it.

That flexibility is not something that the U.N. is likely to afford us. Nonetheless, the U.N. has for many years been offering or advertising itself as a center for solutions to the problem—if such it is—of global warming. And it is the U.N. which to a significant extent has been driving the policy end of the climate change debate for a long time.

How the United Nations Works

The U.N. often is judged by its stated intentions: to promote prosperity, peace, and so on. But it is useful, in many cases, to look not just at the U.N.'s stated intentions but at its incentives. In my experience, those are more powerful as a predictive model of how the U.N. will behave than are its stated intentions.

What is the United Nations? It is a collective. It is run by a grand committee of 192 member states, which means that in theory everyone is responsible—but in practice no one is responsible. If the U.N. pursues policies that encourage its own staff, member states or affiliates and "partners" to fabricate, skew, cheat, bribe, or lie, there are no strong incentives for any one member state to stand up and blow the whistle.

In David Henderson's presentation today, you heard a classic example of everybody piling on, of no one really going against the tide, in the IPCC process. The IPCC is a U.N. organization, and as such it is effectively unaccountable. The U.N. is an organization that operates outside the law. It is subject to no single jurisdiction. Look at where the secretariats shown in Figure 1 are based—in places ranging from Nairobi to New York to Bonn to Geneva, and beyond.

This global web, coupled with U.N. immunity, has been a problem, not least, for U.S. federal prosecutors in New York, who have been trying in recent years to get a handle on a wide range of U.N. misconduct—including not only cases related to Oil-for-Food, but also bribery, kickbacks, and money laundering linked to the U.N. Procurement Department. Again and again, these prosecutors run into obstacles such as Swiss banking secrecy, or inability to gain access to records of front companies in places such as Cyprus, or to extradite former U.N. staff or others indicted in U.N.-related misconduct. A similar murkiness applies to cases involving not allegations of criminal behavior, but skewed data, fiddled policy prescriptions, and U.N. programs designed with built-in perverse incentives.

So who is there actually to enforce anything, to ask hard questions, to hold the U.N. accountable? The answer is nobody. More and more, we have been hearing—even from the U.N.'s own staff—that the U.N. is held to no impartial and systematic external standard of responsibility. A recent study commissioned by the U.N.'s staff union, conducted by a panel of highly respected experts on law and judicial systems, found that the U.N. has no reliable internal mechanisms for holding its own actions to account, either. Thus, we have a large, amorphous, unaccountable institution which, at this point, has found ways to draw funding in amounts that greatly exceed the assessed dues of member states. These days, the U.N. taps into funding sources ranging from private individual donations—like the UNICEF can that your kids shake at Halloween—to corporate donations to fundraising mechanisms that involve, say, charging fees at the World Intellectual Property Organization, to trust funds set up by the hundreds, collecting money from it's-not-clear-whom, for projects which are not spelled out to the public in any detail, with no public accounting of exactly how the money is spent, no clear bottom line, and no mechanisms that ensure effective oversight and clear responsibility within the U.N. itself.

For a sample of just one of the opaque ways in which U.N. funding often works, take the governments of Scandinavia. They field only a few seats in the General Assembly, and they do not have large clout in the U.N. Security Council, where they have no permanent seat. (The permanent five members of the Security Council are the United States, the United Kingdom, China, Russia, and France). But the Scandinavian governments have apparently figured out that they can wield enormous influence in select cases by putting large amounts of money at the U.N.'s disposal via trust funds, leveraged via the U.N. logo, and dedicated to their pet projects. They do this through U.N. agencies, such as the U.N.'s flagship agency, the U.N. Development Program (UNDP), where they can pick the projects that they want to sponsor. In recent years this has led to such disturbing oddities as, for instance, a Scandinavian-funded UNDP project bankrolling the foreign travels and "research" of a number of North Korean officials tied to Pyongyang's weapons programs, under the label of educating these North Korean officials in the art of disarmament negotiations. (This came to light only as a result of

107

the larger "Cash-for-Kim" scandal, in which the UNDP was discovered to have been funneling money to North Korea for everything from round-trip business class air tickets for its officials to attend U.N. board meetings in New York to cash payments never fully explained, as well as the curious feature that the UNDP office in Pyongyang had been keeping counterfeit $100 bills in its office safe.)

These non-appropriated funding sources are all leveraged by the diplomatically immune, largely unaccountable U.N. We need to remember that governments are run by people who sometimes have their own agendas beyond the official agenda, and the U.N. is no different. Its senior officials essentially get to select many of the things that they want to do.

The Oil-for-Food Program and Its Aftermath

We have seen U.N.-related scandals erupt in North Korea and Burma (Myanmar) recently that reflected exactly the operations of the U.N. mechanisms that I have just described. The mother of all recent U.N. scandals was the Iraq Oil-for-Food program during the Saddam Hussein regime. Oil-for-Food in theory was supposed to be a vast humanitarian endeavor. But for the U.N. itself, Oil-for-Food served in many ways as just another fundraising mechanism—albeit especially enormous.

Oil-for-Food was many things, but officially it was a program that began operations in 1996 to provide relief to the people of Iraq while they were living under U.N. sanctions against their government. To administer the program, the U.N. collected a commission of 2.2 percent on Saddam Hussein's oil sales. The bigger the Oil-for-Food program became, the more the U.N. collected in what were in effect commissions on Iraqi oil sales. Not surprisingly, Secretary-General Kofi Annan pushed repeatedly to expand the size and scope of Oil-for-Food. What began as a limited, ad hoc, temporary relief arrangement turned into one in which the U.N. presided over more than $110 billion worth of oil sales plus goods purchases by the government of Saddam Hussein. Over the six years that the program was in operation this structure, together with enabling a global web of sanctions-busting graft, brought some $1.4 billion into the U.N.'s coffers.

How does all this off-budget fundraising compare with the U.N. budget? That is a harder quarter to answer than one might suppose. The U.N.'s system is such a labyrinth that we do not actually know the size of the U.N. budget. In fact, one of the complaints of congressmen looking into it two years ago was that they could not figure out how big the U.N. budget actually was. It is officially something like $2 billion a year for the Secretariat, but let us remember that the Secretariat is just that office in New York. The U.N. system is huge, with many agencies, many more than you can readily calculate a budget for.

Kofi Annan, toward the end of his tenure, finally disclosed an estimated total U.N. budget, without a clear breakdown, of $20 billion a year. In that context, the $1.4 billion received from Saddam Hussein via the Oil-for-Food program represented a fair chunk of change. It funded a lot of U.N. jobs. It was the source of a lot of patronage. We should remember that government bureaucracies like that sort of thing.

When the United States and its allies invaded Iraq in 2003, the records that came tumbling out of Baghdad raised serious questions. It turned out that the U.N. had been presiding over a monumental scam.

Billions of dollars were had been grafted out of the U.N. Oil-for-Food program. Saddam had maintained secret bank accounts abroad for his illicit profits and had used the money to fund everything from palaces to weapons to terrorist groups (we can debate the links to Al Qaeda, but Saddam's funding for Palestinian terrorism was clear). Dozens of countries were involved, and thousands of companies. Every fraudulent device you can imagine—under-pricing oil exports, over-invoicing imports, front companies for shipments that never existed but were billed for anyway—evidence of all this activity spilled out.

Very little of the wrongdoing under Oil-for-Food has been prosecuted, except in the United States. Most governments just chose to look the other way. Again, this is not surprising because some foreign government officials appear to have been actively complicit in the scams.

What we should have understood from all this background material is the potential at the U.N. for something comparably scandalous to build up over time to dimensions that, in some ways, would far exceed even Oil-for-Food. But who is watching out for such a scandal to appear? There is no single dedicated press corps that covers the U.N.'s operations worldwide. National press corps tend to cover their own countries' interests in U.N. actions and little else. No single policing agency or judiciary had jurisdiction over all aspects of Oil-for-Food, for example, so we had a stunning scandal, the upshot of which is nothing more than that a handful of people have gone to jail, or are going to jail, in the United States.

I believe that the scandal occurred in no small part because the U.N. had every incentive to continue the Oil-for-Food program and tried to do so because it was collecting a commission on the oil sales. The commission money went for U.N. jobs. The jobs involved all sorts of budget-padding, as we later learned from the Paul Volcker Committee's inquiry and from other sources; then Saddam was overthrown, and the financial bonanza that was Oil-for-Food was taken away from the U.N.

What the United Nations Learned from Oil-for-Food

One way of considering our discussion today regarding the U.N. is that the U.N. learned an operational template from the Oil-for-Food experience. That is, the U.N. bureaucracy learned that programs like Oil-for-Food are a way of tapping into global big business, into the main currents of the world economy, in a way that brings the U.N. bureaucracy all sorts of employment and benefits.

And among global industries generating big money, few can rival the energy business. Indeed, what business lies at the core of the climate change debate? Energy. Everybody, or so it seems, is now talking about what? Deciding who can burn how much fuel.

In a world that embraces the U.N. line on climate change, there is a lot of easy money to be made by institutions large and powerful enough to act as the gatekeeper, the tollgate, the place through which one has to pass in deciding what is distributed and how. We are actually discussing here, in this conference, a political agenda involving an enormous and global reallocation of resources. In that process, the U.N. apparatus has found its opening for becoming that gatekeeper.

In the last century, this kind of thinking and decision-making was called central planning. It was discredited under that name by the collapse of the former Soviet Union. So we no longer refer to "central planning" when it reappears in other contexts. Instead, we now talk about trading schemes and other allocation devices that sound as though they involve markets. But somebody has to decide who gets what at the end of the day.

If we actually come to having the U.N. try to regulate the climate of the world based on the variables that the scientists here are telling us about, U.N. officials might think that they can understand and foresee some of those things. But they cannot be sure what to make of them entirely, and U.N. efforts to manage the climate of the world would be subject to all the

110

failings that I have just described. In such a scenario, we could have in the making a scandal that someday would make Oil-for-Food look like a drop in the ocean.

How to Create a New U.N. Agency

At this point, I need to go somewhat beyond the IPCC to talk about some of the other U.N. agencies because the tale of how the global environmental project actually evolved at the U.N. is not a simple story. For schemes and intrigue, it more closely resembles something like the Roman imperial court depicted in the public television series, *I, Claudius.*

The U.N., in some sense, is a multi-billion-dollar global think tank or talking shop (with no accountability and very long lunch breaks) where the bureaucrats can come up with almost any project, and if they can find funding for it, they will go forward. Thus, the same U.N. that has proved incapable of auditing its own books now aspires to regulate the weather of the world.

Here is the U.N.'s recipe. If you are a U.N. bureaucrat, or a government official advocating a particular project for the U.N. to undertake, you call a conference. To hold the conference, you set up a secretariat at the U.N., and that can be two people in a room with a telephone.

Once you have the secretariat, you have a permanent foothold. Bureaucracies do not tend to shrink. They tend to become bigger as long as they can keep the money coming in. Again, one of the problems of making the U.N. accountable is that there is no one who walks in, as the voters do in this country occasionally, and says, "Wait a minute. That is my tax money that you are proposing to spend."

An individual American taxpayer is at too distant a remove from what actually flows into the U.N. The amount is seen in Washington as inconsequential spending. After all, the whole U.N. budget is supposed to be only $20 billion a year, and the United States contributes a little more than $5 billion of that. That is chicken feed compared with the size of the federal budget, now around $3 trillion per year.

However, the U.S. contribution to the U.N. is chicken feed that flows through an important corridor—mainlined via the governments of the world. Leverage is the name of this game. Where $5 billion might not get you very far if you wanted to go drilling wells or building bridges privately and directly, $20 billion that flows constantly through an organization that consists

111

of governments will buy you all sorts of leverage for your projects.

Next, after your special U.N.-sponsored conference is held, some new entity is set up, and you wind up with something like the United Nations Environmental Program (UNEP). That is just one of the environmentally related U.N. agencies.

Everything on the chart in Figure 2 represents many offices with many jobs. The UNEP is in Nairobi, Kenya. There is no effective oversight at all. Nobody sends the press the internal audit reports. The internal audit reports of the many U.N. agencies are not even available to the member governments that are supposed to oversee the general bodies that govern the agencies.

In private business, the lack of audit accountability and oversight would be unthinkable. But inside the U.N., those involved cannot see what is wrong with this situation.

Any one of the entities in Figure 1 could be examined the same way, and its organization chart would proliferate into branches like those of the UNEP. There are so many agencies and branches now, in fact, that there is a mind-numbing duplication—and replication—of activities. Remember also that, as Professor Wunsch said, once you are dealing with the climate, you are dealing with the science of everything. Thus, you can find ways in which every agency at the U.N. can become involved.

How the U.N. Uses Carbon Emissions as a Measure of Virtue

An agency called U.N.-Habitat declaims constantly about the effect of cities on climate, and it has produced numerous studies on that point. In a world where low per capita carbon production is a measure of virtue, the World Bank has begun to put out lists of data on carbon emissions for all its member states.

I looked at one of those World Bank lists recently. As those lists are interpreted at the U.N., a country's environmental virtue is assessed largely through the prism of climate change. In this light, what matters is not whether a nation's government is a tyranny or a democracy, a cause of human misery or a servant of human progress. What matters is a bureaucratically derived estimate of carbon emissions.

The carbon emissions charts show that one of the real champions of minimizing carbon output is Cuba at 2.3 metric tons per capita. Zimbabwe is even better at 1.2. Of course, in Zimbabwe, that carbon output is so low

Figure 2. The United Nations Environmental Program (UNEP) Organization Chart.

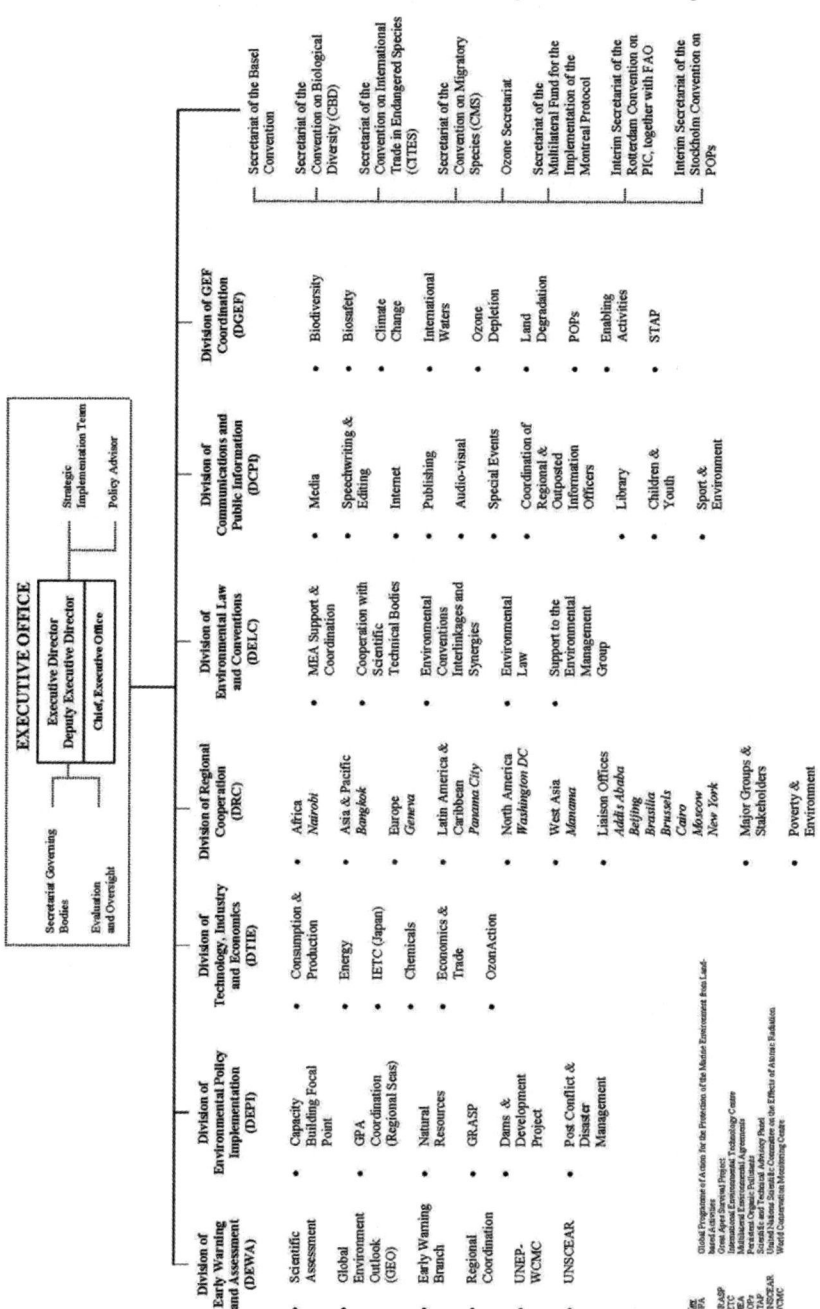

because the economy has declined to the point where people are starving.

Equatorial Guinea has one of the world's most repressive regimes and is a major oil exporter but is a real low-carbon-output star at 0.4 metric tons per capita. The People's Democratic Republic of Laos has similarly low emissions. China, which is big and messy and dirty, still has carbon output of only 2.2 metric tons per capita because most of China is still dirt-poor.

North Korea's output is 3.5 metric tons per capita. I suspect that the government there fiddles with its numbers to make them look as though they are more productive than they really are. (North Korea illustrates some of the U.N.'s more bizarre priorities. Via the UNDP, the Global Environment Facility was used in the 1990s to bring funding into North Korea to pay for preserving biodiversity at a mountain that happens to be a favorite haunt of the Party elite. While the U.N. was preserving that "biodiversity," an estimated one to two million people in North Korea were dying of famine.)

Iran currently is one of the world's great sources of tension, threats, and certified terrorism-sponsoring, meanwhile seeking the nuclear bomb. But you know the story: 4.6 metric tons of carbon emissions per capita—not bad.

Finally, we come to the real villain on the list: 20.5 metric tons per capita for the United States. But what the U.N. climate-change mindset does not consider is that we are also one of the most productive, creative, and inventive nations. We contribute to medical discoveries that lengthen human life. America has given the world inventions like the automobile and the airplane—an astounding array of inventions that actually allow people to live better.

U.N. Supervision of Trading of Carbon Offsets

I could, with tongue in cheek, suggest that if we are going to have carbon offsets traded through the U.N., we should be allowed to have corruption offsets also, because corruption taints the international system and costs people in many ways. Should we not also have despotism offsets because despots, in fact, do terrible things to the world order? You can make the case that the actions of despots result in higher infant mortality in places where people live less well.

Regarding offsets for population growth, we know through studies over the last 20 years or so by economists like Nobel laureate Gary Becker that, as people become wealthier, they tend to have fewer children. People make tradeoffs involving how they invest their time, and the palatable answer to how to deal with population growth is to find ways for people to become richer.

There are whole sets of offsets that we could have if we actually start down this path because climate is not our only international concern. Terrorism offsets would be a good idea: States that sponsor terrorism, which imposes all sorts of costs on us, via everything from airport security screening lines to worry, concern, and the potential destruction of property and lives, should pay a penalty for that sponsorship.

But the U.N., with its despot-packed General Assembly, is not about to institute the trading of offsets such as those that I have just proposed. The weather is a much safer topic of discussion, one that does not really antagonize member states, except for the guilt-stricken, wealthy, productive countries of the world.

The trading of carbon offsets suggests that consideration be given to trading waste offsets generally. One of the hallmarks of that list of desperately poor countries that I presented above, which ranked so high on the carbon emissions honor list, is that they have governments that completely stifle the opportunities of their people. Those populations otherwise would be capable of living quite as well as we do.

The problem is that it is not actually ordinary people who tend to profit from trading schemes of the types that actually are likely to be implemented. It will be their governments instead, in most instances.

Again, we have to remember that the U.N. is an organization of governments. We tend to think of it as a place that may represent people because we vote for our government and because the U.N. engages in voting procedures. But the Chinese are not well represented by the Chinese government at the U.N., and neither are the North Koreans or the Iranians, and so forth.

Sources and Consequences of the U.N. Fixation on Climate Change

Let us now discuss how the current U.N. fixation on climate occurred. Is it not strange, after all, that out of all the things for which you could trade offsets—corruption, despotism, terrorism, warmongering, and other really bad ideas fed into the world system—it is climate that has become such a prime concern?

The first explanation for the U.N.'s fixation on climate change is that it has great potential for generating U.N. income as the U.N. delves more deeply into the business of regulating it. This is all about the big bucks energy business.

Another explanation is that, in the history of politics, both foreign and domestic, it is convenient to be a person with a plan. Please allow me to give you one interesting case study of someone who came to the U.N. a long time ago with a plan. In 1947, he worked as a security officer in the nascent U.N. headquarters: Maurice Strong—he is best known these days as the godfather of the 1997 Kyoto Treaty.

From his humble entry into the U.N. system 60 years ago, he went on to perform various jobs at the U.N., and over the years he became one of the gray eminences behind the scenes. He was the head of the secretariat that arranged the Stockholm environmental conference in 1972. After that conference, he arranged to have UNEP set up in Nairobi. He then ran UNEP himself for the first three years. In 1992, he presided over the U.N.'s Earth Summit at the Rio conference. He later set up or helped set up and run various non-governmental organizations that became intertwined with the U.N. while serving as an adviser to the World Bank, to former U.N. Secretary-General Boutros Boutros-Ghali, and until recently as a top aide to Kofi Annan.

At present, Mr. Strong (a Canadian citizen) is in China. Although he frequently worked at U.N. headquarters in New York, he appears to have been staying away from the United States since information surfaced in the Oil-for-Food investigations in 2005 to the effect that, while working for the U.N. in 1997, he received a check for $988,885 from a bagman since convicted of conspiring to bribe U.N. officials on behalf of the government of Saddam Hussein. (Strong has not been accused of any wrongdoing. He initially told investigators that he did not recall receiving the check. When they showed him the canceled check, with his signature, he said that he had been unaware that the money originated in Baghdad—and thought the purveyor merely wished to invest in a Strong family company.)

At the U.N., Strong's persistence over many years was a major factor in setting up the universe of U.N. environmental agencies, departments, and campaigns that exist today. It was through the UNEP (which he set up and ran) and the World Meteorological Organization that the IPCC emerged, as David Henderson has noted. Around the time of the Rio conference, followed five years later by Kyoto, Strong already was at work on a pilot project called at the time the Global Environment Facility (GEF) to be headquartered in Washington.

That GEF is now a joint activity of UNEP, the UNDP (where Strong served as an adviser), and the World Bank (where Strong also served as an adviser). As you can see, these bureaucratic structures replicate and replicate: Arrange a conference, set up a secretariat, start a new agency, and soon you have another permanent office under U.N. auspices to serve your cause.

Strong is someone who learned early on how to work the U.N.'s system, and he has been pushing his particular agenda for a long time, trying to make money off private ventures along the way. The U.N. system is not configured to deflect gross conflicts of interest, or even to recognize them, for that matter.

Here is how gross those conflicts can become. It turned out, during the Oil-for-Food investigations, that Maurice Strong had been brought into the Secretariat by Kofi Annan in 1997 to do sweeping reforms of the U.N. system.

Among those sweeping reforms, he created the Oil-for-Food "Office of the Iraq Programme" inside the U.N. Secretariat that employed the now-fugitive-on-Cyprus Benon Sevan, who ran the Oil-for-Food program. Two weeks after the creation of that office was announced, Strong received that $988,885 check—for what he said was a private business deal—issued from a Jordanian bank, funded by a bag of cash sent from Baghdad.

Setting aside the question of where, exactly, that money came from, there is still the question of whether the U.N. should have permitted Strong—while he was engineering huge changes in the structure of the U.N.—to accept million-dollar checks from anybody. Of course, he was working at the time for Kofi Annan, who in 2006 personally accepted a $500,000 cash prize from the ruler of Dubai for work on the environment (Annan finally relinquished that prize, weeks later, after the story broke in the press that he had appointed as the new head of UNEP a member of the jury that awarded him the prize).

The tale of Maurice Strong is a sample of the kind of activity that happens beneath the surface at the U.N. Meanwhile, on the U.N. main stage, there are lofty discussions about the implications of climate change for the planet.

U.N. Operations in Geneva, Switzerland

The headquarters of the World Meteorological Organization (WMO) are beautiful, housed in a high-tech building with an excellent cafeteria on top and huge windows looking out on the Swiss scenery. Inside, the scene is a bit murkier. Swiss authorities have been investigating alleged financial deals involving a friend and business partner of Kofi Annan's son, a man named Michael Wilson, and a real estate swap connected with another U.N. agency, the World Intellectual Property Organization (where there has been yet another big flap recently, because the Director General turns out to have falsified his own date of birth).

The WMO also was the home, for many years, of a Sudanese employee named Mohammed Hassan, now suspected of having filched millions out of WMO fellowship-for-training programs. When an auditor began digging into the fellowship programs, Hassan fled the country, apparently faked his own death, and remains missing. The WMO fired the auditor.

U.N. Prize Money

Regarding the IPCC, Walker Todd suggested that I find out what it was going to do with the IPCC's share of the 2007 Nobel Peace Prize money (about $900,000). I called and asked. They do not know yet. They need to decide at the IPCC meeting in Valencia, Spain, later this month (November 2007). That is interesting because the IPCC is a governmental panel. But it does not know what it is going to do with the prize money.

(Author's Note: As of March 31, 2008, an IPCC spokeswoman, reached by phone in Geneva, said that the IPCC still has not decided what to do with the prize money.)

U.N. Administration of Carbon Trading

The UNDP is now trying to launch a carbon emissions trading facility linked to a cap-and-trade regime. Even if the world does decide to go forward with such cap-and-trade schemes, one might wonder why the U.N. should be entrusted with managing any part of it.

Is there some reason why the U.N., which cannot audit its own books and accounts to no one, should be in the business of overseeing the trading of any financial contract? The only reason I can think of for doing so is that the trading would be yet another spigot of money—unaccountable, badly audited money—pouring into the U.N. system. That is quite an inducement to put climate change high on the list of U.N. interests and to promote it as a priority regardless of any evidence that dissenting scientists might discover. Every U.N. agency wants a cut of the climate action—many are already collecting funds and handing out the per diem allowances.

A recent development is that Kofi Annan's successor as Secretary General, Ban Ki-moon of Korea, has extended climate change into traditional diplomatic fields, doing things like writing in *The Washington Post* that the real underlying trouble in Sudan is climate change.

For the U.N., the focus on climate change is a bonanza. The UNEP alone has a budget of $132 million a year, paid for by governments around the world. It replicates and replicates, because that is what the U.N., in fact, is best set up to do. For the rest of us, I would call a U.N.-managed approach to climate change the perfect storm.

CAP-AND-TRADE, OR CARBON TAX?

Kenneth P. Green

B ECAUSE these days, anyone who utters a negative sentiment about a climate change proposal is labeled a "denier," I should like to start with a brief description of my scientific and philosophical beliefs. Having trained in the environmental sciences, I studied the underlying literature on climate change as a doctoral student in 1990. Ever since then, I have believed that:

- There has been a broad global warming since 1850, by about one degree centigrade;
- Some of that warming is very likely attributable to human greenhouse gas emissions and land-use changes, though how much is still unclear; and
- If it continues, such warming poses a likely risk to future generations.

Having worked as a policy analyst and lay economist for about 15 years now, I also believe that global warming is one of the most challenging policy issues that we ever have faced. After all, the topic is all about energy, and energy is central to economic development. Because of that focus, our policy approach is absolutely crucial, or we risk easily doing ourselves and future generations far more harm than good.

Unfortunately, climate policy has been mired in flawed approaches since 1994, when the issue of climate change was inextricably bonded to the general values of the United Nations in the United Nations Framework Convention on Climate Change. As Claudia Rosett's presentation made vivid for us, it is fair to say that the U.N. is overwhelmingly bureaucratic and technocratic in its approaches; explicitly redistributionist; prone to being jerked around by developing countries; largely anti-capitalist; and in many ways hostile to what are called the "Western powers." The Kyoto Protocol, reflecting all these values, essentially was doomed to fail. Kyoto Round Two is equally likely to fail, being grounded in precisely the same set of values.

While U.N. studies have paid lip service to the need to adapt, or to learn to sequester carbon, the amounts of research money, lobbying money, and academic laurels pointed in those directions have been trivial in comparison to what was poured into the push for the initial Kyoto Protocol and its sequels.

We still, I should point out, pay far too little attention to adaptation and sequestration and too much attention to mitigating greenhouse gas emissions. Our reaction now is like what a short-sighted person might do when told that over-watering the lawn has cracked the foundation of our house. We have focused entirely on cutting back our watering while ignoring the spreading cracks in the foundation.

Mitigation Policy Choices: Cap-and-Trade versus Carbon Taxes

Cap-and-trade, the leading policy contender, is favored by environmental groups and those generally on the left of the political spectrum. I argue that it is fundamentally flawed for reasons similar to those for which Kyoto was flawed: it is technocratic, bureaucratic, anti-capitalist, and redistributionist.

Carbon emissions taxes, by contrast, are embraced by a growing band of economists and policy analysts from across the political spectrum, including many on the libertarian right.

The next part of my presentation is based on an Environmental Policy Outlook that several coauthors and I published at AEI this past June.[1]

Both theory and practice tell us that cap-and-trade would be generally ineffective in reducing greenhouse gas emissions. The European experience in this regard is informative, with virtually all the countries engaged in the European carbon emissions trading system on track to miss their emission reduction targets by a wide margin.

A cap-and-trade system causes increases in the prices of several things. Energy prices increase, which is the main point of the exercise. A cap on carbon emissions amounts to a cap on supply in the face of rising demand. Increased energy prices thus become unavoidable.

The prices of goods and services increase. Energy being a fundamental input for economic productivity, everything dependent on energy inputs increases in cost as well.

Energy price volatility increases. The business cycle and many other factors guarantee alternating periods of sharp economic growth and slower growth. As energy demand increases with economic growth, periods of growth lead to price spikes in energy costs as demand shoots past supply.

Non-Price Effects of Cap-and-Trade Schemes

Emissions trading schemes tend to be highly prone to fraud. Parties involved in trading have incentive to cheat on historic baseline estimates, to claim credit for things they would have done anyway, and to exaggerate the benefits of carbon reduction technology. Carbon dioxide emissions are extremely hard to measure locally; thus progress is based on self-reported reductions in emissions. Credits generated by offsets are particularly hard to validate because one cannot know how long an offset will remain in existence or how effective it will be over the long term. Many countries selling offsets have lax property rights. Governments wanting to claim credit, and lacking revenue or tools for enforcement, have the incentive to wink and nod in the face of such cheating. Finally, studies of the existing offset market, offering a taste of things to come, have shown that market to be largely fraudulent.

Cap-and-trade tends to be massively redistributionist and regressive, raising the costs of energy asymmetrically (e.g., coal prices have increased more than natural gas recently), thereby harming some parts of the country (the coal-consuming states) and benefitting others (natural gas-consuming states). Higher energy costs also tend to hit the poor more heavily than the better-off.

The effectiveness of cap-and-trade schemes can be negated by policy safety valves. Most cap-and-trade schemes being floated have either overt or covert safety valves to allow covered entities to continue emitting if the cost of reduction exceeds a given price (usually about $15 per ton of carbon dioxide). Such safety valves render the cap fictitious and the trading element of the scheme irrelevant.

Anti-Competitive Effects of Cap-and-Trade

Cap-and-trade schemes tend to create massive new national and international bureaucracies. At present, there are no institutions set up to oversee a national trading regime in carbon emissions, much less an international one.

The resulting corporate and governmental bureaucracies involved in cap-and-trade will tend to become self-entrenching. Once a group holds permits worth a significant amount of money, the group has a vested interest in perpetuation of the system and the prevention of alternative mitigation approaches that might devalue their carbon currency. This has been called the "carbon cartel" problem.

There is a self-tightening effect in cap-and-trade systems. Major holders and sellers of carbon permits would see nothing but benefits in ratcheting the caps ever more tightly. They may be expected logically to lobby accordingly.

Errors in establishing or policing cap-and-trade systems might prove irremediable. Though permits are not described as true property rights in the literature, it is inconceivable that a trading system could be discontinued without buying out the permit holders or otherwise indemnifying them against losses. Given the trillions of dollars that ultimately could be involved, the trading system will be as immune to repeal as any major entitlement program.

An important consideration is that a cap-and-trade system usually will create no revenue for enforcement or to offset economic damage caused by emission caps. Auctioning initial permit rights, the rational and economically most favored solution to this problem, never has happened and never will happen in a meaningful way. Most cap-and-trade schemes auction off a pitifully small fraction of the initial permit allocations.

Price Effects of a Revenue-Neutral Carbon Emissions Tax

A revenue-neutral carbon tax, by contrast, would be generally effective in reducing greenhouse gas emissions. The increased prices inevitably would suppress demand and reduce possibilities for fraud, thus insuring that emission reductions would be real.

A carbon tax would increase energy prices, which is the entire point of either system: to suppress energy demand through pricing carbon emissions. In turn, the increased energy prices would increase prices of goods and services generally, which is the same effect as with cap-and-trade.

However, in contrast to cap-and-trade, a carbon tax would tend to be price-stabilizing. The larger the share of energy costs that one pays in taxes, the less an underlying fluctuation in energy prices would change the final price to consumers.

A carbon tax would be less redistributionist. The absence of a trading scheme means that coal-consuming states would not have to ship revenues to natural gas/nuclear-power/hydro-power states to buy emission permits. Yes, there would be disproportionate impacts, but they would not be accompanied by redistribution of wealth as well.

Non-Price Effects of Carbon Taxation

Corruption is a lesser possibility with carbon taxation. Given that a carbon tax would raise a massive revenue stream for the government, the government certainly would have the incentive for enforcement that would be missing with cap-and-trade. Also, the government could use existing collection mechanisms for the carbon tax. For better or worse, the Internal Revenue Service and state sales tax collection agencies are always with us.

A carbon tax would allow for regulatory streamlining. A vast number of existing regulations become redundant with a carbon tax, such as corporate average fuel economy (CAFE) standards, appliance standards, construction standards, light bulb standards, and so forth.

A revenue stream would be created that could be used to offset economic damage from the tax. The importance of that revenue stream for protecting economic growth cannot be overstated. There is, of course, the eternal threat that politicians would waste the revenue or funnel it into pet projects, but that same risk would exist even if permits were auctioned off in a cap-and-trade system.

Carbon taxes can be adjustable in the face of changing conditions. Tax reform happens on national election cycles—compare that process with international protocols like the Kyoto negotiations, which can take a decade (or more) for all parties to agree on changes.

Conclusion: Environmental and Economic Benefits from a Carbon Tax

Modeling calculations that my coauthors and I performed indicated that a tax of $15 per ton of CO_2 emitted would reduce the use of coal by 83 percent, oil by 11 percent, and natural gas by 9.6 percent.[2] Such a tax would add about $0.14 to the price of a gallon of gasoline. The tax would raise about $80 billion annually, which could be used to reduce individual income taxes by 13 percent, corporate income taxes by 29 percent, or payroll taxes by 10 percent. In virtually all respects, a revenue-neutral carbon tax is a superior policy to carbon cap-and-trade systems.

Endnotes

[1] Kenneth P. Green, Steven F. Hayward, and Kevin A. Hassett, "Climate Change: Caps vs. Taxes," Environmental Policy Outlook, June 1, 2007,

American Enterprise Institute. Available online at: http://www.aei.org/ publications/filter.all,pubID.26286/pub_detail.asp.

2 Green et al. (2007).

GLOBAL WARMING AND RELIGION:
CLIMATE POLICY AS APPLIED THEOLOGY

Robert H. Nelson

IT has long been predicted that a warming of the earth, to the extent it occurred due to increasing greenhouse gases, would have the greatest effects at night, in the winter, and towards the poles. These are the times and places where cold weather is experienced at its fiercest. Throughout human history, the cold was something to be survived and avoided. Only the use of fire, shelter, and clothing made it possible to live in the coldest climates. It might thus seem that a warming of the earth's climate specifically occurring mostly in the night, in the winter, and toward the poles would be a welcome development. Of course, we know that for many people this is not the case. It is interesting to ask why.

This summer offered a small case study in the Arctic. As ice receded in the summer according to the longstanding pattern, the area of open sea in the Arctic Ocean this time was much larger than seen before (since satellite observations started in 1979), suggesting that a warming climate was already beginning to have some clear observable consequences for the earth's environment. As *The New York Times* reported, advocates of stronger measures to limit climate change promptly "cited the [Arctic] meltdown as proof that human activities are propelling a slide toward climate calamity."[1]

As far as direct impacts on human beings, the increasing area of open sea of the Arctic hardly seems calamitous. Sea level rise would be minimal because the Arctic is ice and no land; the new volume of water released by melting would approximate the former volume of ice. Some other consequences seemed positively heartening. Christopher Columbus discovered the Americas in searching for an ocean passage to India and many other explorers followed. Now, a northwest passage might at last be at hand, shortening travel times and otherwise cutting shipping costs between the east coast of the United States and China and other Asian countries. As the Panama canal has become overloaded and antiquated for many purposes, there have been discussions of building a new canal in Central America that would cost in the tens of billions of dollars. Spending all that money now might not be necessary.

Yet more economically consequential, the Arctic may contain trillions of dollars worth of oil and gas reserves. With the opening of Arctic seas, this potentially immense petroleum wealth might become newly accessible to human discovery and exploitation. Indeed, concerns are now being raised that there are no well-defined property or other international systems of rights to define the possession of Arctic petroleum reserves. Russia has been making aggressive claims and statements with respect to its expansive interpretation of its own Arctic domain. Canada, Norway, Denmark (Greenland) and Finland, also potential Arctic claimants, are observing these developments warily, along with the United States. There might be the possibility of a real calamity here but not the one raised by environmental observers; in the worst case, there could even be a threat of armed conflict. Nations have often fought in the past over control of scarce natural resources—the Middle East now an example.

When the critics of current climate policies go beyond generalizations, the one "calamity" to which they specifically point in the Arctic is the fate of the polar bears. There are at present about 22,000 polar bears in the Arctic regions—sufficient numbers that polar bears as of this writing are not listed as a threatened or endangered species. The precise impact on polar bear populations of shrinking Arctic ice fields in the summer is difficult to predict but some investigators suggest it could be a loss of up to two-thirds of the bears by the mid twenty-first century—and in the worst case this might happen even more precipitously.[2] Aside from the general biodiversity considerations, the one direct economic impact on human beings would be a loss of hunting—about 400 polar bears a year are now hunted. In short, the negative direct consequences for human economic welfare of shrinking Artic ice in the summer would seem to be small.

Moreover, I submit that—and while no "scientific" proof can be advanced—the current situation in the Arctic is broadly applicable to climate change and global warming across the world as a whole.* Human beings have spread over the earth because they are the most adaptable species.

---

* The *Stern Review on the Economics of Climate Change*, officially released by the British government in November 2006, argued otherwise but its conclusions— much in contrast to most previous economic studies— were based on extreme and inappropriate assumptions. See William D. Nordhaus, "A Review of the *Stern Review on the Economics of Climate Change*," 45 *Journal of Economic Literature* 3 (September 2007). In many books and articles, a leading student of the economic impacts of climate change, Robert Mendelsohn of the Yale School of Forestry and Environmental Studies, has in fact argued that much of

At least within the range of warming considered most likely for the next 100 years, the adaptations required by global warming would probably be significant but not fundamentally different in degree from other adaptations within recorded human history. Indeed, they would probably be considerably less than required in the past to deal with the consequences of the single most disruptive force in human history — warfare among nations, tribes, and other groupings internal to the human species. Thus, for the next 100 years at least (time that can be spent developing new and superior adaptive strategies), the economic consequences of global warming for the world as a whole should be well within the capacity of human beings to deal with them. One can never be sure but it does not seem likely that there is any major world "calamity" in prospect – rather, large scale but hardly catastrophic economic readjustment.

Like the polar bears in the Arctic, however, many individual plant and animal species have weaker adaptive capacities. Global warming holds out the prospect of a large scale biodiversity reshuffling for the earth. At least some species will go extinct. At least some other species will find that their competitive evolutionary prospects are enhanced. For a large number of people, like the changes now taking place in the Arctic, all this is seen as ecologically "calamitous" on a worldwide scale.

But why? If it is not at heart because of the economic consequences for human beings, and it is certainly not a strictly scientific conclusion (science being a method of inquiry that is value-neutral in such matters), why would a large change in the ecological order of the earth be regarded as calamitous? The answer requires a moral judgment. Many people believe it would be morally wrong — a virtual evil to put it bluntly — for human beings to change the ecology of the earth in this drastic manner. But, then again, why? Why is this powerful moral judgment being made with respect to a changed earth?

Addressing this issue might seem to be a question for philosophy. But contemporary philosophers typically strive to be analytical and non-judgmental.

---

the temperate zone of the northern hemisphere would in fact experience economic benefits from a modestly warmer climate. Empirically, since the development of air conditioning, the largest population movements in the United States have been from colder to warmer parts of the nation, suggesting a preference for at least a somewhat warmer climate. See Robert Mendelsohn, "The Peculiar Economics of Global Warming," *The Milken Institute Review* (Second Quarter, 2000), 31-37.

Few today undertake to articulate and defend modern versions of the Ten Commandments. It is necessary, I submit, to turn to religion. Religions, unlike philosophy, have not abandoned the role to make and defend strong moral judgments of good and evil. Moreover, some of the most influential moral judgments today—such as the "sinful" character of human actions that are changing the climate of the earth—are being made by "secular religions" such as environmentalism.* The study of contemporary religion thus must extend broadly to include the theological grounds for these secular moral judgments.[3]

The most important secular religions of at least the past 150 years have been Marxism, the American progressive "gospel of efficiency," social Darwinism, and still other forms of "economic religion."[4] Economic religion sees economic progress as the path of the salvation of the world. There is little basis, however, in economic religion for seeing recent developments in the Arctic, or with respect to the earth's climate as a whole, as a deeply offensive event. A large reordering of the earth's ecological regimes is not in itself a great concern within the framework of economic religion. Economics focuses on human welfare for which, as noted, the prospects seem reasonably good, as long as the economic freedom to make appropriate economic adjustments is preserved.

Rather, the fundamental objection to climate change is not to the economic consequences but to the very fact of the change itself—to the biological implications for the many plant and animal species such as polar bears in and of themselves. It is a moral objection in principle to any large scale human reordering of the ecological workings of the earth.† This is the province not of economic religion but of environmental religion.[5] More specifically, it is the focus of attention for a new gospel of conservation biology. Despite

---

* Steven Rockefeller and John Elder state that: "The global environmental crisis, which threatens not only the future of human civilization but all life on earth, is fundamentally a moral and religious problem." See Steven C. Rockefeller and John C. Elder, eds., *Spirit and Nature: Why the Environment is a Religious Issue* (Boston: Beacon Press,1992), p. 1

† In 1992, Al Gore warned that "artificial global warming" was looming and "it threatens to destroy the climate equilibrium we have know for the entire history of human civilization. As the climate pattern begins to change, so too do the movements of the wind and rain, the floods and droughts, the grasslands and deserts, the insects and weeds, the feasts and famines, the seasons of peace and war"—all such changes greatly for the worse. See Al Gore, *Earth in the Balance* (Boston, MA: Houghton Mifflin, 1992), p. 98

the many contemporary assertions to the contrary, climate change is not a calamitous economic threat to the human species as a whole. But it is a large biological threat to many of the other plant and animal species of the earth. Assessing the moral significance of changing the ecology of the earth is not a matter for economics, science, or any other professional expertise but for religion—and theology is the systematic exploration of the beliefs and grounds for religion.

I. The Gospel According to Conservation Biology

The Society for Conservation Biology was founded in 1985 and its influential journal, *Conservation Biology*, was established soon thereafter in 1987.[6] Subsequently, programs for the study of conservation biology were created at a number of leading American universities. The field of conservation biology has become the focal point for considering changes in the ecological order and the policy measures that should be undertaken in response—and climate change offers the single largest human action potentially affecting the ecology of the earth on a worldwide scale.

In 1992, environmental historian David Takacs—who describes himself as a "lifelong environmentalist"—interviewed 23 leading figures in conservation biology, including Michael Soule, Reed Noss, E. O. Wilson, Thomas Lovejoy, Paul Ehrlich, and Jerry Franklin. His purpose was to explore their reasons for becoming involved in the field, the methods of conservation biology, the values reflected in its efforts, the prospects for the future, and many other matters. His findings are directly applicable to climate change as the single most important policy area affecting the future biological state of the earth. Takacs assembled the materials from these interviews, conducted additional interviews specifically concerned with environmental issues in Costa Rica, and in 1996 published an insightful book, *The Idea of Biodiversity: Philosophies of Paradise*.[7] Although Takacs did not use this characterization, it was a theological inquiry into the new secular religion of conservation biology.

Takacs sees the rise of conservation biology, and the new focus on a goal of biodiversity, as a reflection in part of perceived problems with earlier environmental goals, especially the wilderness ideal of protecting wild nature that is seen as little touched by human hand. There was a growing awareness by the 1980s that pre-modern human impacts on the natural world might have been much greater than previously suspected. In Costa Rica, for example, Takacs noted that "researchers are turning up pottery shards and crop residues that point to past civilizations where until recently we had imagined only wilderness." More broadly, an environmental goal of "wilderness preservation . . . is redolent of class privilege, culturally rooted, and ontologically precarious." In light of these and other concerns, by the 1980s there was a new perception that "plotting conservation around wilderness is a dubious strategy."[8]

The Idea of Biodiversity

As Takacs explains, the questioning of a wilderness strategy was one element in a broader concern among biologists and other scientists with respect to "the negative connotations the word *nature* holds." Nature was the subject of romantic poetry, transcendental philosophical speculations, and many other approaches that fell well short of the "scientific." If environmental goals were to gain wider public acceptance, it might be desirable to put them on a firmer scientific footing. Indeed, this was a main purpose of elevating the goal of "biodiversity" in place of the older and now seemingly less compelling environmental language of "nature, wilderness, natural variety, endangered species, and biological diversity."[9]

A central issue explored by Takacs in his interviews with conservation biologists was the definition of "biodiversity." An expert in ecological processes, Don Falk, considered that biodiversity takes in "ecosystem functions, community processes, genetic diversity within species, and so on." Given such a broad scope, Takacs sought to examine how the idea of biodiversity might differ from the idea of nature (that is, what is separate from human activity). He concluded that for most conservation biologists there was in fact not much difference. The language of biodiversity mostly amounted to a rhetorical act of "scientizing the concept of nature."[10] A similar observation might be made with respect to efforts of current climate activists to portray the consequences of global warming as an inevitable "calamity" for human beings—thus camouflaging their real moral judgments that concern the impacts on nature in and of themselves.*

---

* As Mark Sagoff has observed, "ecology in large part has become the science of Eden." There is an underlying belief that "Nature has ecological integrity and design because it is directed by an independent Force"—that is to say, by God. It has become the case today that "it is our ecologists and philosophers who now impute overarching order, purpose, or design to the natural world." In other words, the science of ecology offers new metaphors by which environmental religion speaks to modern men and women who otherwise reject

Whether conservation biologists consciously intended this or not, their goals were actually political and social. As Takacs explains, "the word *biodiversity* is part of a convincing strategy—that is, it is designed to convince [the American public and political leadership] and has been quite convincing thus far" in terms of advancing traditional environmental purposes in a new language. Takacs summarizes the forces at play in the establishment of the new field of conservation biology (and that are now being applied in parallel ways by scientific activists in the area of climate change):

> Conservation biologists do not often go to bat for nature per se; they do not often describe nature in their writing. According to Neal Evernden, "The environmental advocate sits on the horns of a dilemma: the time honoured technique of invoking the authority of nature has been essential to the presentation of a persuasive argument, and yet that technique is now vulnerable to charges of fraud." The term *nature* not only carries a multiplicity of confusing, often self-serving meanings; it also carries the taint of association with bleeding-heart liberal tree huggers. To be considered a "nature-lover" is not a compliment in many quarters. So rather than running to nature, biologists flee from it. Instead, they describe and defend biodiversity. It maintains a scientific aura of respectability while still meaning so many different things to so many different people, without having yet acquired the notorious etymological reputation of the word *nature*. [11]

The leadership of the field of conservation biology has come predominately from biological scientists. Unlike many scientists who are content to work in their laboratories, however, and to leave politics and policy making to others, most conservation biologists—like a large number of climate change scientists—have been determined to make a difference in the world.

Conservation Biology as Crusade

Takacs also explores the roots of the crusading spirit that has animated the efforts of so many conservation biologists—and now so many climate

---

religion in its older forms. See Mark Sagoff, "Ecosystem Design in Historical and Philosophical Context," in David Pimentel, Laura Westra, and Reed F. Noss, eds., *Ecological Integrity: Integrating Environment, Conservation, and Health* (Washington, DC: Island Press, 2000), p. 74-75 change activists who share the concern for the biological order of the earth. Not surprisingly, the sources lie outside the scientific method and the biological expertise possessed by the participants. Indeed, despite their efforts to distinguish themselves, conservation biologists are sustained by powerful ethical ideas and spiritual values similar to those of the older conservation and environmental movements that sought to protect "nature." Most advocates for the biological order, however, have not been particularly self-reflective about all this; they have not applied a scientific and analytical lens to explore the contents of their own powerful value feelings associated with biodiversity (and nature).

This probably reflects in part the ambivalent feelings that might be aroused. Conservation biologists (like many climate change activists) advertise themselves as belonging to the scientific community and as adhering to strict cannons of scientific objectivity and value-neutrality. As Takacs notes, however, with respect to conservation biologists, in actual fact they "attempt to speak for values that go far beyond what one might think of as falling within their realm of expertise." They engage in "public advocacy" in support of a powerful "ethical imperative," one that encompasses many "extrascientific values."[12]

Some of Takacs's conservation biology interviewees were more self-aware than others in recognizing the tensions between their public advocacy and their strictly scientific roles. This tension showed up in the advice given to some young biologists without tenure that they might need to "wait before they engage in conservation activities as part of their professional lives." Walter Rosen, who originally came up with the term "biodiversity," stated his concern that "science is supposed to be objective, yet I, who am a scientist, nevertheless feel very strongly in this and that value." Moreover, there could even be an element of misrepresentation, Rosen acknowledged, because "if I'm going to be listened to, it's probably because I'm a scientist, even though I'm making a non-scientific assertion."[13] Takacs suggests that one solution might be for conservation biologists to distinguish clearly in their public roles between their statements as scientists and their statements as citizens of the world who are advocating particular values and biodiversity policies. But then he acknowledges that in practice any such attempt to separate these two roles is probably unworkable. One might say the same of a number of leading scientists, such as James Hansen, in the climate field.

Substantial portions of Takacs's book are devoted to exploring the contents of the powerful value systems that he finds underlying the public advocacy of conservation biologists—now extended by climate change activists to the ecology of the whole earth. A number of conservation biologists agreed that biodiversity is so important to the world because of its "transformative value," a concept first developed by the environmental philosopher Bryan Norton. When a person is "surrounded by diversity," there is an identification "with the natural world; one is inextricably part of it. The transformation of values occurs partly because if you are inextricable from the grand process of nature, by consuming it or altering it, you irrevocably hurt yourself."[14] However large or small the economic impacts might be, it follows that changing the world's climate by human actions would also be damaging to the future "transformative" power of nature throughout the world, a consequence transcending those human actions in any one localized area.

Analogies to religion came readily to mind when conservation biologists spoke of the transformative power of experiencing biodiversity. Takacs comments that they "seek to encourage this 'conversion effect' by putting people in direct contact with biodiversity. Biologists may feel such conversion is possible because they themselves went through precisely this kind of transformation, usually in childhood." Indeed, there are parallels here to being born again in Christianity. E. O. Wilson relates that for him "natural history came like salvation at a very early age." Another conservation biologist Thomas Eisner describes a youth in which he was "exposed to the smell of the woods, to looking under rocks and looking under logs. And there was just an overwhelming feeling." Reed Noss expresses his sense that "many people do have that feeling, that there is a larger self. And when they're defending nature, they're defending that larger self."[15] Climate change activists experience similar feelings in their defense of the whole earth's existing natural biological order.

A transformation is something that a person can describe as an actual event or experience that either has or has not occurred in their life. Takacs also examines the importance of "intrinsic value" in the thinking of many conservation biologists. Given that intrinsic value by definition exists "apart from any human valuer," it is not an observable and measurable event and, as Takacs notes, it goes "well beyond the realm of what we might expect scientists to acknowledge and defend." Indeed, intrinsic value may have to be justified from "certain religious standpoints. If God or some other

135

deity or sacred process created the natural world alongside humans, then all creatures are imbued with sacredness: all have intrinsic value" independent of any human thoughts or actions.[16]

Thus, one way to understand how intrinsic value might be outside any human scope is that it is really the value that God has given to "His Creation." When they talk about intrinsic value, many conservationist biologists (and climate change activists) may in essence be talking about following the commands of a Jewish and Christian God, even if mostly without realizing it. Paul Ehrlich, one of the conservation biologists who is committed to the idea of intrinsic value, does not ground his arguments in Genesis but he does go so far as to recognize explicitly that "this is fundamentally a religious argument. There is no scientific way to 'prove' that nonhuman organisms . . . have a right to exist." Takacs notes the irony that many conservation biologists proclaim beliefs dependent on the existence of a God or other deity and yet "most biologists have no such religious views" that they can articulate in any detail.[17]

At one point in his interview process, Takacs raised the subject of religion more explicitly, asking specifically about the role of "spiritual values" in the thinking of conservation biologists. This is a difficult area for many conservation biologists because "if it seems a priori odd that some scientists believe and preach a concept like intrinsic value that cannot be proven scientifically—indeed, it can barely be expressed at all—it may seem totally bizarre that scientists talk about biodiversity's spiritual value." Yet, perhaps the majority of the conservation biologists interviewed spoke in terms of having deep spiritual convictions relating to biodiversity. S. J. McNaughton described his powerful "spiritual experience" in once being surrounded by wildebeests and other nature on an African plain. Reed Noss described his strong sense of "a kind of spiritual or at least a nonrational connection to nature." Noss hastened to add, however, that "I wouldn't call it religious." Takacs comments that many conservation biologists make similar distinctions, reflecting the fact that, among his interviewees, almost all "these biologists reject organized Western religion, sometimes quite forcefully."[18] Thus, they are willing to admit to having a strong sense of "spirituality" in the presence of nature while rejecting the idea of following any institutional Christian or other forms of Western "religion."

While conservation biologists may not have a systematic theology, it was nevertheless evident to Takacs that ideas and reactions of a deeply religious character were central to the whole enterprise of conservation biology, and now it appears extending to the climate change activists who are so fearful of the worldwide impacts of global warming for the earth's biodiversity,

> Such feelings run deep, infusing their bearers with sentiment. At a loss for language adequate to express this sentiment, they resort to the word one resorts to when one can't explain something: *spiritual*. For these biologists and for many others, being in nature—surrounding oneself with biodiversity—can almost not help but bring about experiences to leave the senses reeling, the mouth agape. The incomprehensible complexity of it all: we can't handle it. Our brains go numb when faced with such richness out there, so much bigger than ourselves. How can we help but feel awed? And biologists spend their lives digging deeper into the intricacies, developing profound awareness of both the mindblowing intricacies they have unearthed and the complicated skein they haven't begun to entangle. [19]

Although such religious experiences are widespread among conservation biologists, Takacs notes that many are reluctant to "speak out publicly because they feel they must preserve the boundaries between rational and intuitive, mind and body, science and emotion." Crudely put, putting their intense religious feelings about biodiversity into the public view might blow their scientific cover. For climate change activists, similarly, admitting that worldwide biodiversity—"the ecological state of the earth"—is the real issue, and that direct human welfare in an economic sense is a side question (and may not even be greatly affected), would have similar consequences. A few conservation biologists, admittedly, do think that this reluctance is a mistake; for one thing, the cause of conservation biology, like the strongest measures to limit climate change, probably cannot succeed unless the core values are more widely adopted among Americans, and this process of conversion will require a more explicit statement of their religious significance. Since "the values are there already," Takacs says, "why not be honest, making conservation biologists' work more accurate and holistic? Simultaneously, they'd be laying their values bare for others to emulate."[20] For one thing, there would be the possibility of a powerful political alliance with devout Christians who share much the same sense of awe and reverence in the presence of nature—for them, God's Creation.

In any case, there was no mistaking the fact that, among the conservation biologists Takacs interviewed, they "attach the label *spiritual* to deep, driving feelings they can't understand, but that give their lives meaning, impel their professional activities, and make them ardent conservationists. Getting to know biodiversity better takes the place of getting to know God better." Indeed, despite the reluctance of most conservation biologists to use the term "religion" in describing their own beliefs, this was mainly an act of linguistic camouflage. It was quite evident to Takacs that "some biologists have found their own brand of religion, and it's based on biodiversity." It might pose a threat of some outside critics speaking harshly of conservation biologists as being the new "eco-ayatollahs," but Takacs argues that conservation biology should now more courageously and honestly put its true religious face forward—as perhaps should climate change activists today.[21]

II. How Much Is God Worth?

There might be another possibility, however. What if professional economists could put a dollar figure on the intrinsic value of—the simple fact of the very existence of—a species, forest, climate, or other object in a "natural" condition? What if the existing state of the world's biodiversity has an economic value for its own sake, independent of any direct human uses? If such estimates could be made, they could then be factored into overall economic calculations, and government decisions concerning changes in the climate and other human impacts on the natural world could still be made as the result of an economic benefit-cost analysis. Economics would encompass the existence of biodiversity, a species, a natural climate, and other ecological values as part of the full domain of economic calculation.[22]

This would admittedly require a large departure from economics as traditionally practiced. Human beings, the way of thinking of economics has long assumed, live for happiness and happiness is a result of direct consumption alone. As the economist Stanley Lebergott says, "the goal of every economy is to provide consumption. So economists of all persuasions have agreed, from Smith and Mill to Keynes, Tobin, and Becker."[23] Historically, there has been no place in economic thinking for the idea that something that is never seen, touched or otherwise experienced—that is not consumed in any way by any person—can have an economic value to an individual (or to society).

Yet, the economic way of thinking was at odds with the emerging environmental ethos of the 1960s and 1970s—and now of the increasingly influential field of conservation biology. Instead of a "natural resource" to be used to increase the production—and ultimately the consumption—of goods and services, the natural world should be protected for its own sake, even if there is no direct human benefit. While obviously not a practical possibility, one environmentalist went so far as to assert that "every intervention in nature that cannot be rectified is a sacrilege," a concept of the intrinsic value of existing nature that historically appears nowhere in economics. [24]

But, in 1967, a leading environmental economist at a Washington research institute, Resources for the Future, proposed a solution. John Krutilla suggested that the scope of economics should expand to include a new concept, which has since come to be known as "existence value." [25] Formally, the "variables" in a person's "utility function" would include not only the amounts of food, clothing, and other ordinary goods and services consumed but also the various states of knowledge that each person has of the existence of plant, animal, and other natural features in the world—now including the existence of animal species such as polar bears in the Arctic. Implicitly at least, consumers would be willing to pay something for this latter form of "consumption," the existence of a pleasant thought in their mind, thus giving rise to efforts by economists to calculate existence values in precise dollar terms.[26] Preserving the whole earth's climate in its current state would presumably have one of the highest existence values that had ever been a possible subject for economic calculation, potentially an enormous economic benefit involving the continued existence of the states of nature all over the world that could then be factored into climate policy debates.

By the 1980s, the concept of existence value was coming into use by a number of environmental economists for purposes such as estimating the benefits of government actions or calculating damage assessments against business corporations whose actions had harmed parts of the natural world.[27] A federal appeals court in 1989 directed the Department of the Interior to give greater weight to existence values in its assessment of money payments for damages to public resources under the Superfund law.[28] The concept even received favorable notice in literary circles such as the *New York Review of Books*, where the author of one article argued that it should be central to achieving world biodiversity objectives: "But why should citizens of industrialized countries pay to preserve resources that are legally the domain of other countries? An obscure tenet of economics provides a rationale. Certain things have what is known as an 'existence value.'"[29] More recently, an

MIT economist finds that the economic profession is concerned with studying three types of value — "use value, option value, and existence value." Although there may be many conceptual issues to resolve, economists can not continue to ignore the fact that "consumers may want [that] some good or state of the world continue to exist, and be willing to pay something to sustain it, even though they know they will never experience it directly, or 'consume' it in any normal sense."[30]

A Growing Debate

When the idea of calculating existence values was raised, most mainstream economists at first paid little attention. However, as the potential uses widened and as the policy stakes escalated, an active professional debate broke out by the early 1990s.[31] Non-economists also entered the controversy.[32] As some suggested, it appeared that environmental economists were attempting to calculate the economic value of obeying an ethical command such as God's instructions to Noah in the Bible. Many environmentalists recognized the potential political gains from showing high dollar estimates of existence values but otherwise found the concept offensive to their own environmental beliefs.

The Exxon Corporation, facing large potential damage assessments as a result of the Exxon Valdez oil spill in Alaska, and believing that these assessments might be based in part on existence value estimates for the loss of natural features in Prince William Sound, committed large financial resources to studying the issue. Exxon hired a number of leading economists — many of whom were not environmental economists and were new to the subject — to examine whether the calculation of existence values was an appropriate economic method. Their critique was on the whole negative.[33] The State of Alaska and the federal government hired several leading environmental economists who took a more positive view.[34]

Hoping to resolve the issue, the National Oceanographic and Atmospheric Administration (NOAA) convened a panel of leading economists, chaired by economics Nobel prize winners Kenneth Arrow and Robert Solow. In 1993, the NOAA panel declared that, although there were significant practical problems in estimating existence values and much care must be exercised to prevent their misuse, the concept should in principle be incorporated into the set of economic tools.[35] However, an active debate has since continued.[36] As economist Henry Jacoby commented in 2002, when it comes to the use of "techniques of contingent evaluation" (in estimating existence values), the majority of economists did not have much confidence in them at present—even as they were reluctant to concede that important domains of social choice fell outside the realm of legitimate economic calculation.[37]

Thus far, those who have actually attempted to measure existence values have studied mostly wilderness areas, threatened species and other environmental concerns. However, the use of the concept is potentially much wider. There appear to have been few if any attempts to calculate the dollar value of maintaining the existing climate of the earth (and thus also the existing natural order that is sustained at current temperatures) but this is an obvious candidate. There are also many nonenvironmental existence values. Tropical forests may have an existence value but there will also be an existence value in knowing of the higher incomes of poor people in less developed countries—some of whose jobs may depend on cutting the forests.

Indeed, there is virtually an endless set of possibilities for the calculation of existence values.[38] Almost any state of the world invested with symbolic significance by large numbers of people will have a non-trivial existence value. For example, the very presence of an abortion clinic in a community will cause some of the residents of the community to feel psychic pain, while others will feel good about it. An act of burning the American flag will have a large negative existence value for many patriotic citizens, although a large positive value for others who may "enjoy" the symbolic assertion of free speech rights in American political life.

Once the full scope of existence values was recognized, extending well beyond environmental concerns, economists were in effect proposing to substitute formal economic calculation as a way of resolving such basic value controversies. Many essentially ethical and moral questions were to be resolved by public answers to economic survey instruments, asking how much was the dollar worth to a person of one mental image versus another. And it was not merely a few environmental economists but a number of leading members of the economics profession, including Nobel prize winners, who were advocating this professional course of action.

Finding God's Hand in Nature

In the early 1970s, the *New Yorker* writer John McPhee traveled with the former executive director of the Sierra Club, David Brower, who had for many years been touring lecture halls on college campuses and other places across the United States preaching his environmental message. As McPhee discovered, there were environmental prophets, great texts, sacred sites. According to McPhee, "throughout the sermon, Brower quotes the gospel—the gospel according to John Muir, . . . the gospel according to Henry David Thoreau."[39] Brower was a direct follower in the line of Muir, who had founded the Sierra Club in 1892.

For Muir the wilderness had an explicitly religious significance. He referred to primitive forests as "temples" and to trees as "psalm singing." As Roderick Nash wrote in *Wilderness and the American Mind*, Muir considered that the "wilderness glowed, to be sure, only for those who approached it on a higher spiritual plane. . . . In this condition he believed life's inner harmonies, fundamental truths of existence, stood out in bold relief."[40] This reflected Muir's belief that in the natural objects of wild areas it was possible to find "terrestrial manifestations of God." They provided a "window opening into heaven, a mirror reflecting the Creator," making it possible to encounter in nature some true "sparks of the Divine Soul."[41]

In creating the world in six days, God had demonstrated his divine workmanship that was now available to instruct human beings in His ways. It was in the limited areas of wilderness that still remained, as Muir considered, that one could find places still genuinely "steeped with God"—where the Creation still existed in the original form.[42] Yet, as a result of the spread of science and industry in the modern era, this available opening to the mind of God was being erased. Human beings were building dams, cutting forests, farming the land and in any number of other ways—now including perhaps the most destructive human act of all, changing the climate of the earth—were imposing a strong human footprint on the Creation. If at some point in the future all the wild areas were lost, future generations would be forever cut off from this main way of knowing about God.

All this is to say that for Muir a wilderness area was literally a kind of church. A church is a place of spiritual inspiration. It is a place where people come to learn about and to better understand the meaning of God in their life. It is above all in church settings that God communicates his messages for the world. A wilderness church, furthermore, is in one sense more imposing and more awe inspiring than any human-built church. In environmental religion, a wilderness is a church literally built by God. In nature untouched by human hand, environmentalism says—if often only in an implicit fashion—that it is possible to directly experience the Creation.

The Arctic covers a larger area than most wildernesses but its great environmental importance lies in the fact that it has been among the wildest places of the earth, the least impacted by human actions—but even the Arctic wildness is now being altered by climate change.

Today, these religious convictions that inspired Muir often still lie behind the protection of wilderness areas. Roger Kennedy, while he was serving as the director of the National Park Service in the Clinton administration, stated that "wilderness is a religious concept" and that "we should conceive of wilderness as part of our religious life." Wilderness puts us "in the presence of the unknowable and the uncontrollable before which all humans stand in awe"—that is to say, although Kennedy did not put it in just these words, in wilderness we stand in the presence of an actual work of God.[43]

How Much Is a Church Worth?

If the concept of existence value were extended into every possible area of life, it would be possible to ask: how much is believing in God—having the idea of God firmly planted in one's mind—worth in dollar terms? What is the intrinsic value of knowing about God and His ways? Calculating a monetary value for the existence of a wilderness area, however, comes close to the same thing. Nature untouched by human hand, as found in a wilderness area, is for the John Muirs of the world—many of them still members of the Sierra Club—a means of obtaining greater knowledge of God. In contemporary environmentalism this message comes in only a slightly revised form—explicit mention of God is admittedly left out but wild nature is "the true source of values for the world," or other such environmental rephrasings of traditional religious messages.

Admittedly, to value a wilderness in this way is to value the instrument of communication of religious truth rather than the actual knowledge of God. Thus, a more precise analogy would be: How much is the existence of a church worth—including the benefits to those who to come to the church and find their faith reaffirmed and also to those people who will never visit this church but are happy to know of its existence? This is, at least in concept, an answerable question. Economists can point out that, although leaders of organized religions may be offended by the question, they do in fact make such calculations. Other things equal, more churches are likely to be better. But more churches also cost more money. In making a decision at some point to build another church or not, a religious organization is in effect saying that the money benefit of the additional church is or is not worth the cost of building and maintaining it. However crass it may seem to say, on the margin the additional worth of greater communication of God's thinking to the world does not create a dollar benefit large enough to cover the added dollar expenses of one more church.

So how would one go about putting a marginal value on the existence of one more church (one more wilderness)? Answering this question, assuming a person is willing to think about the matter in these economic terms, would involve multiple concerns. One question to be addressed would be: How much does a particular new church (wilderness) add to the religious education of the faithful? How many new people might it draw into the faith? Related to this would be the question, how many churches (wildernesses) should a religious denomination ideally maintain and how many does it already have? This obviously depends partly on the total number of faithful, their geographic distribution, and the expected growth of the religious group in the future.

Yet another factor, to be sure, is that the building of a church is not just a way to be spiritually uplifted. It can also be a way of publicly and symbolically announcing a depth of religious commitment, a way of formally taking an action for the greater glory of God. Building a grand cathedral, such as Notre Dame in Paris, can take on a special religious significance when it involves a great sacrifice of effort—as religions have historically found meaning in making large sacrifices of many kinds to express their deep devotion to their gods. Many primitive tribes, for instance, have had religious rites in which they might sacrifice their best cow or goat in order to honor their god.

A wilderness area thus might become all the more meaningful in much the same way: The more valuable the mineral, timber, and other natural resources given up, the greater is the sacrifice made and the greater the symbolic statement of allegiance to the faith. Indeed, this may be why the Arctic National Wildlife Refuge (ANWR) in northeast Alaska has become so important to the environmental movement. It is not just the on-the-ground environmental features of the area—in both Alaska and Canada there are in truth many other equally desolate and isolated places bordering on the Arctic. The truly distinctive feature of ANWR is that so much valuable oil and natural gas—estimated amounts that would have a worth of more than $800 billion in gross dollar value at current oil prices—would potentially

144

be sacrificed. At these prices the net income (after accounting for costs) of oil and gas development of ANWR would likely exceed $600 billion. If this area is instead left "untouched," the ANWR "church" conceivably would be the most expensive cathedral ever dedicated to any faith in the history of the world, forever redounding to the glory of god, and the American environmental movement.

The potential economic sacrifices involved in actions to prevent world climate change, to be sure, would dwarf the economic sacrifices in leaving ANWR undeveloped. With respect to climate change, the stakes are much higher because the negative impacts on the existing natural order would be pervasive throughout the world. Although there is no one identifiable area that would amount to an environmental church—unless one considers that the entire earth is such a church—the idea of a halting climate change as a sacrifice to demonstrate commitment to environmental values (devotion to an environmental god) is still applicable. It might also be in part a form of dollar restitution for our past economic sins against nature. This is another "benefit" that would have to be factored into a comprehensive economic calculation of the net social value of alternative climate change policies.

In considering proposals by economists to measure dollar existence values—including the monetary value of maintaining the existing world climate—environmental leaders do in fact commonly react much as other religious leaders would. While recognizing a potential gain in political support in putting their case in conventional economic terms, environmentalists have on the whole been cool if not antagonistic to efforts by economists to calculate existence values. Mark Sagoff, a past president of the International Society of Environmental Ethics, writes that the calculation of existence values is "an attempt to expand economic theory to cover environmental values. . . . But what makes environmental values important—what makes them values—often has little or nothing to do with 'preferences,' with perceived well-being, or with the [utilitarian] 'satisfaction' people may feel in taking principled positions." Aside from the many practical analytical problems, Sagoff rejects existence value in principle as an imperialistic attempt by economists to substitute clever analytical techniques for "the role that the public discussion of values should play in formulating environmental policy" in a democratic society.[44]

As Sagoff and others are saying, a religion must be judged by the validity of its claims to offer "the truth" of the world, not on the basis of the dollar value of the pleasures found in thinking about the religion or about its leading public symbols. Over the course of history, surprisingly many people have even chosen martyrdom over a coerced renunciation of their understanding of religious truth. When religions offer different understandings of the world, the matter can not be settled by dollar calculations of the pleasures or displeasures associated with alternative religious mental imagery. To assert otherwise, as surprisingly many environmental economists were now doing, was simply to make an imperialistic claim for the ultimate validity of economic religion—that the god of economics is the true divinity and that the environmental and other gods are subordinate or false altogether.

III. Environmental Creationism

Most conservation biologists have been brought up in a world in which the western religious tradition still resonates strongly, even when many of its institutional representatives in the temples and churches have been in decline. In the modern age the Christian (and also Jewish) understanding of the world has survived in new forms, sometimes in total unawareness of the original source.* It would seem that the field of conservation biology

---

* Jean Starobinski is a leading interpreter of the Enlightenment thought of Jean Jacques Rousseau. Although many people in the eighteenth century failed to notice, the biblical parallels have become increasingly obvious with hindsight. As Starobinski now comments, Rousseau "takes the religious myth and sets it in historical time." In order "to explain the fall of man [there is] no demon tempter or tempted Eve—no supernatural intervention is required; human causes [alone] will suffice." For Rousseau, as a result of misguided human actions in the world, there has been an "ever-increasing burden of human artifice" that in its consequences has "accelerated the fall into corruption: such is the history of mankind." The disastrous outcome has been a loss of that original "state of nature [in which] man lived happily in peace. Appearance and reality [then] were in perfect equilibrium. Men showed themselves and were seen by others as they really were. External appearances were not obstacles but faithful mirrors, wherein mind met mind in perfect harmony." But all was lost in the path to the modern obsession with progress. Rousseau had rewritten the first book of Genesis but in a naturalist language more suited to the mood of the eighteenth century—and, as it would turn out, also better suited for many others in the nineteenth and twentieth centuries, including many current environmentalists.

It was not only Rousseau who remapped the Christian message to secular vocabularies. As Starobinski comments, intellectual historians of the twentieth century increasingly recognized that in the Enlightenment "the philosophes' major ideas are for the most part secularized religious concepts." See Jean Starobinski, *Jean-Jacques Rousseau: Transparency and Obstruction* (Chicago, IL: University of Chicago Press, 1988—first French edition, 1971), pp. 12, 112.

is yet another example of this modern phenomenon of powerfully felt and expressed secular religions that operate without the traditional language of religion. While conservation biologists almost all reject Christian creationism, they may not have travelled as far as they are accustomed to thinking. The descriptions they give of experiencing biodiversity are little different from the classic Christian feelings of religious awe and reverence in experiencing the presence of "the Creation."

Indeed, it would seem that, like so many earlier ecologists and other environmental advocates for the protection of nature such as John Muir, most conservation biologists are also believers in one or another form of "environmental creationism." It is the experience of the creation, inspiring a deeply felt sense of encountering a true work of God, and coming to a greater understanding of God's thinking and His design for the world, that is so enrapturing to conservation biologists. No other explanation can adequately account for the intensity of their religious feelings in the presence of biodiversity. These feelings certainly do not come from "value-neutral" science. Nor can science account for the intense feelings of those who fear that climate change is altering the ecological order of the plant and animal species across the whole world.

Playing God Environmentally

The lack of theological introspection among conservation biologists in some cases has had the consequence of ensnaring them in a tangle of contradictions.[45] In both the Christian religion and the gospel of conservation biology, the greatest sin is to play God in the world—as human beings are now doing with respect to the world's climate. Yet, conservation biologists also strongly advocate the contemporary environmental goal of restoring a state of the natural world as it existed before significant human alteration. Moreover, this goal has been widely adopted as a matter of government policy. Restoration of nature lies at the core of the ideas of "ecosystem management," now the official management philosophy of the U.S. Forest Service, National Park Service, Bureau of Land Management and many other government organizations that deal with the natural world.[46] In the western United States, where European settlement mostly did not occur until after 1850, restoration has often been defined operationally by these public land agencies to mean the recovery of the "pre-European" condition of the lands and other features of the environment.

147

The popularity of the restoration agenda has reflected its underlying creationism. If seldom stated this explicitly, both conservation biologists and the American public have in the back of their minds that restoration will successfully accomplish an act of "recreating the creation." The problem here, however, is that this would seem to be yet another act of playing God—in this case environmentally. The restoration goal assumes that human beings have the same knowledge and the same capacity to recreate the natural world that God possessed in the beginning—in other words, if the restoration is really genuine, they would be seeking to become virtual new gods themselves.

Aside from this theological quandary, conservation biologists seem to be thinking of a static world, dating from the creation, that can then be restored and appreciated for its wonders in its original form. It is a way of thinking surprisingly similar—especially in light of their past mutual hostility—to the thinking of traditional Christian creationism. We see here once again the close affinity of conservation biology and Christian theology, the former providing perhaps a disguised but scientifically more acceptable way of delivering important messages of the latter.

Yet, at the same time, and seemingly unaware of any possible contradictions, conservation biologists advocate a strictly Darwinian understanding of the world, one with which Christian creationists have waged fierce battles. In a dynamic evolutionary world, moreover, what would be the meaning and purpose of restoration? What would be the restoration target point in time, even assuming the technical capacity to achieve it fully? Any particular choice of a target would seem rather arbitrary. While government agencies in the western United States have settled upon 1870 to 1890 as the moment when humans in effect began to play God there (the point at which Europeans arrived with their extraordinary scientific and economic powers), some people have suggested logically enough that the appropriate moment of original "true nature" in the West must have been much further back, preceding the arrival of native Americans as well.

Even assuming a time frame could somehow be specified, any objective to restore the creation would be a mere fantasy. In an evolutionary world, it will be impossible to single out any one past moment of the evolutionary process to declare that this was the actual moment of "the creation." Whatever it might be called, and the terms used by conservationists biologists and environmentalists are numerous – including a uniquely "natural,"

148

"healthy," "sustainable," "equilibrium," "biologically diverse," and so forth particular state of the world—it will be merely a fiction of the creation. Thus, even if it were somehow miraculously possible to restore a precise natural moment of the past, the value of this "fake" human version of the creation would not be very great.[47]

Schizophrenic Scientists

Nevertheless, large amounts of money are already being spent in the United States to "restore" the natural environment—and these levels of funding could well increase substantially in the future. Much larger spending, of course, is involved in efforts to maintain the existing state of the world climate—and in the future, if warming can be arrested, these efforts might then be directed at restoring the climate to the current world temperatures. It is unclear, to be sure, what is actually being "restored"—although we can say for sure that it is not the creation. In practice, restoration money will probably be spent to remove dams, canals, trails, bridges, power lines, and many other symbols of the past scientific management of nature for economic purposes. Formerly drained wetlands will be reflooded, rivers will be returned to their former channels, and many other similar steps will be taken. Symbolically, it will be a repudiation of our past false worship of the god of economic religion, the deity in whose name many of these physical manipulations of nature were mistakenly undertaken, brazenly challenging God's authority. As far as the eventual environmental outcome on the ground, it is likely to be something that is brand new. Most of the heroic "restoration" efforts now being made will probably never result in a natural condition in the world that has ever previously existed before.

Christian creationists have a well developed and internally consistent way of understanding the arrival in the world and the religious meaning of the creation. The problem for them is that the traditional biblical story is contradicted by Darwinism and other products of modern geological and biological scientific investigations. The conservation biologists who advance an environmental creationism experience virtually the same sense of religious awe and inspiration in the presence of nature—the creation—as their Christian counterparts. The many scientific advocates for halting climate change also are motivated by a strong desire to maintain the current state of nature—now encompassing the existing climate of the entire earth. Yet, many of these biological and climate change activists

149

also profess to be true scientists who look to Darwin and other modern science to understand nature.

The result, however, would seem to be almost a schizophrenia in their thinking about the natural world. Many scientists experience nature, on the one hand, as "the creation," and yet also write professionally and talk about it in biological terms as the Darwinian product of billions of years of random mutation and other evolutionary workings. These tensions within the scientific enterprise come to the fore when it is necessary to consider the contents of environmental policies in practice. What is so sacred about the current climate regime? Why would it seem offensive to so many people to attempt to calculate an "optimal temperature" of the earth? Why is there so little scientific interest in and effort devoted to the possibility of geo-engineering of the earth's climate to achieve such a goal? Indeed, even geo-engineering the climate to maintain the current temperatures of the earth is widely seen as very objectionable. The answers to all these questions are not to be found in science but in an implicit environmental religion that drives much of the current thinking about climate change policy—the same religion that is the source of much of the thinking of conservation biologists.

For these conservation biologists, what is it that is being restored? In the event, it cannot be "the creation." It is not necessary, moreover, to restore a process of Darwinian evolution because the workings of evolution never stopped in the world—nor in principle could evolution ever be halted by human action. The goal might be to reset the evolutionary clock to a specific time frame preceding human impacts on the evolutionary result but it now appears that this might be many hundreds of thousands or even a few million years ago and a particular choice would inevitably be arbitrary. Given the theological tenets of the environmental creationism underlying much of conservation biology, in short, the goal of restoring the natural world would seem to be rife with contradictions and incoherent.

Nevertheless, heroic human activities are taking place with a justification of "restoring" nature—and also now of preserving the current climate of the earth. If the results are likely to be problematic, it will be due to intellectual—really theological—confusion as much as any technical or economic considerations. Theology is not just a matter of living a moral life, or finding the path to a salvation in the hereafter. In putting the tenets of conservation biology and climate science into practice, a confused current theological understanding is likely to also produce wide policy and management confusions among the many government agencies that have to address issues of the true and correct relationship of human beings and nature.

Conclusion

In 1849, Prince Albert announced in London a plan for a great new Crystal Palace Exhibition to be held in 1851 to celebrate the extraordinary advances of sciences and industry of recent times. For the Prince, as for many other people of the period, it would herald the fulfilment of God's plan for the world. As the Prince declared, "so man is approaching a more complete fulfilment of that great and sacred mission which he has to perform in this world. His reason being created after the image of God, he has to use it to discover the laws by which the Almighty governs His creation, and, by making these laws his standard of action, to conquer nature to his use—himself a divine instrument."[48] As would have been unimaginable until very recently, the Prince was optimistic that human beings by their God-given powers of scientific knowledge and economic advance would soon be bringing heaven to earth.

One-hundred and fifty years later, matters looked much different. The application of human reason, reaching a high point in the scientific method, had created powers to control nature that could be used for evil as well as good. Reflecting concerns felt around the world, in 2007 a Philippine newspaper columnist declared that: "Man tends to destroy his natural habitat as he consumes its natural resources in his quest for a better life. The worst environmental damages have been results of major manmade atrocities on land or the waters that veer away from what God designed our planet to be."[49] Reflecting a common fear that God would severely punish those who were destroying his Creation, many environmentalists foresaw a future of rising seas, famine, disease, and other natural disasters—environmental calamities virtually biblical in character.

Contemporary economists do not use the Christian language of Prince Albert but they are the heirs to the same belief in scientific and economic progress as the salvation of the world. Environmentalists, by contrast, today reject such progressive religion as a heresy that is threatening the human future on earth. Perhaps the most important religious controversy in the world today is this clash between "economic religion" and "environmental religion."[50] Whether for good or evil, human beings indisputably are altering the natural world in unprecedented ways. The stakes in human actions have been raised as never before. The answers will have to be resolved in the end by theology.

At present, politicians, lobbyists, policy analysts, and other participants in American public debates do have a theology in the back of their minds. But their theological views are largely masked in various pubic disguises. Indeed, the leading religions in public affairs today mostly deny their religious character. This denial, ironically, is sometimes an important part of their theology. Karl Marx once said that religion is the opiate of the masses but it can be seen in hindsight that this outward rejection of religion was actually a main tenet of a new Marxist religion.*

Our leading contemporary theologians thus speak publicly in the languages of economics, natural resource management, conservation biology, ecology, and other forms of official policy discourse. Rather than obeying God's commands, many Americans are more comfortable in communicating their deepest values in languages of technical expertise and rational policy making.[51] Nevertheless, as one American political scientist observed in 2006, it is obvious that "like many other areas of public policy, [environmental policy making] has extremely powerful symbols associated with policy outcomes on all sides." The newer environmental symbols now challenged an older economic set of symbols that venerated "growth and jobs," "private property rights," and other forces of economic progress.[52]

In studying the theological disagreements between economic religion and environmental religion, current analysts thus typically make little or no explicit reference to the "R-word"—religion. About the closest they get is when they speak in terms of the influence on public policy of an "underlying policy frame," a "policy image," or a "dominant causal logic," all of these outlooks on the world resistant to any easy change, even in the face of contradictory empirical evidence.[53] If religion is understood, however, as a person's way of framing his or her basic perception of the world and its meaning, as leading theologians such as Paul Tillich have understood the matter, then these basic perceptual orientations are actually religions in a genuine sense.

----

* The distinguished theologian Paul Tillich once declared that Marx was "the most successful of all theologians since the Reformation." See Paul Tillich, *A History of Christian Thought: From Its Judaic and Hellenistic Origins to Existentialism* (New York: Simon and Schuster, 1967), p. 467.

As commonly studied in schools of theology today, a religion does not necessarily have to involve the existence of a god outside this world.* As many current theologians understand the matter, most people have a set of foundational beliefs to guide their lives and the intellectual articulation of these beliefs in a formal way becomes a theology—whether there is an explicit reference to a god or not. Many people now speak of having deep "spiritual" beliefs while denying that they have any "religious" viewpoint. Indeed, if it makes a reader more comfortable, he or she may prefer to think of the environmental conflicts today as a struggle between the spiritual values of economics versus the spiritual values of environmentalism. For me, though, it is a distinction without a difference.

Partly reflecting the disguised character of the most important modern religions, most of our leading intellectuals have not committed much time and effort to formal theological reflection. Although attitudes have begun to change recently, the greatest efforts of social scientists and policy analysts have been devoted to studying and promoting the mechanisms of scientific and economic "progress." In some private universities, there are schools of theology but they are commonly separated from the main areas of study in the physical sciences, humanities, and social sciences. In public universities, religion can be a legitimate object of "scientific" study but a full range of theological debate is precluded by the "non-sectarian" mission.

Our critical theological debates therefore have been driven underground. The camouflaged form of contemporary theological debate is preferable to no debate at all but the quality of the argument inevitably suffers. There can be large confusions introduced when questions that are essentially re-

---

* Max Stackhouse, professor of theology and public life emeritus at Princeton Theological Seminary, thus declares that "a religion" is a belief system that is "widely shared, it shapes an ethos that gives identify to a particular culture and tends to promote a social ethic that fosters distinctive public institutions." As Stackhouse observes:

> By this definition, worldviews such as a philosophical-ethical Confucianism, an atheistic spirituality such as Buddhism, or a secular-humanistic ideology such as Marxism, whenever they form a creed, a code and a cult, and are used to interpret and guide the formation of an ethos, can properly be seen as faiths. They function as "religions," shaping an ethos, even if they are opposed to theistic traditions or do not recognize themselves as religious. They are also subject to theological analysis, for they inevitably contain a "metaphysical-moral vision"—an ontology, a theory of history and ethic—that involves some view of transcendence.

See Max L. Stackhouse, *Globalization and Grace* (New York: Continuum, 2007), p. 8.

ligious have to be explored in a "value-neutral" language of economics or ecological science—or climate change science. The very claims to value neutrality inhibit closer examination of the powerful normative elements underlying the thinking of those involved. Yet, if these elements remain unacknowledged, conversations between advocates of different viewpoints with respect to the environment—including prominently the state of the climate—will often be like ships passing in the night.

Endnotes

[1] Andrew C. Revkin, "Arctic Melt Unnerves the Experts," *The New York Times*, Science Times, October 2, 2007.

[2] Andrew C. Revkin, "Grim Outlook for Polar Bears," *The New York Times*, Science Times, October 2, 2007.

[3] See Charles Taylor, *A Secular Age* (Cambridge, MA: Harvard University Press, 2007).

[4] See Robert H. Nelson, *Reaching for Heaven on Earth: The Theological Meaning of Economics* (Lanham, MD: Rowman & Littlefield, 1991); and Robert H. Nelson, *Economics as Religion: From Samuelson to Chicago and Beyond* (University Park, PA: Pennsylvania State University Press, 2001).

[5] See Robert H. Nelson, "Unoriginal Sin: The Judeo-Christian Roots of Ecotheology," *Policy Review* (Summer 1990); Robert H. Nelson,"Environmental Calvinism: The Judeo-Christian Roots of Environmental Theology," in Roger E. Meiners and Bruce Yandle, eds., *Taking the Environment Seriously* (Lanham, Md.: Rowman and Littlefield, 1993); Robert H. Nelson, "Sustainability, Efficiency and God: Economic Values and the Sustainability Debate," *Annual Review of Ecology and Systematics*, Volume 26 (1995); Robert H. Nelson, "Calvinism Minus God: Environmental Restoration as a Theological Concept," in L. Anathea Brooks and Stacy D. VanDeveer, eds., *Saving the Seas: Values, Scientists and International Governance* (Maryland Sea Grant College, 1997).

[6] This section of the paper is adapted from Robert H. Nelson, "The Gospel According to Conservation Biology," *Philosophy and Public Policy Quarterly*, Vol. 27, Nos. 3-4 (Summer/Fall 2007).

[7] David Takacs, *The Idea of Biodiversity: Philosophies of Paradise* (Baltimore, MD: Johns Hopkins University Press, 1996).

[8] Ibid., p. 42.

9 Ibid., p. 195.
10 Ibid., pp. 74, 79.
11 Ibid., p. 76.
12 Ibid., p. 115.
13 Ibid., p. 167.
14 Ibid., p. 230.
15 Ibid., pp. 237-239, 243.
16 Ibid., p. 247.
17 Ibid., pp. 248, 247.
18 Ibid., pp. 262, 265.
19 Ibid., pp. 266, 267.
20 Ibid., p. 268.
21 Ibid., pp. 270, 336.
22 This section of the paper is adapted from Robert H. Nelson, "Does 'Existence Value' Exist?: An Essay on Religions, Old and New," *The Independent Review* (March 1997). Reprinted in Robert Higgs and Carl P. Close, eds., *Re-Thinking Green: Alternatives to Environmental Bureaucracy* (Oakland, CA: The Independent Institute, 2005).
23 Stanley Lebergott, "Long-Term Trends in the U.S. Standard of Living," in Julian L. Simon , ed., *The State of Humanity* (Cambridge, MA: Blackwell, 1995), p. 149.
24 J. Moltmann, "The Destruction and Healing of the Earth: Ecology and Theology," in Max Stackhouse and D. Browning, eds., *God and Globalization, vol. 2: Theological Ethics and the Spheres of Life* (Harrisburg, PA: Trinity, 2001), p. 279.
25 John V. Krutilla, "Conservation Reconsidered," 57 *American Economic Review* (September 1967): 777-786.
26 Robert Mitchell and Richard Carson, *Using Surveys to Value Public Goods: The Contingent Valuation Method* (Washington, D.C.: Resources for the Future, 1989).
27 Robert Cameron Mitchell and Richard T. Carson, *Using Surveys to Value Public Goods: The Contingent Valuation Method* (Washington, DC: Resources for the Future, 1989).
28 *State of Ohio v. United States Department of the Interior*, D.C. Circuit. 880 F. 2d 432 (1989).
29 John Terborgh, "A Matter of Life and Death," *New York Review of Books* (November 5, 1992), p. 6.

[30] Henry D. Jacoby, What is Nature Worth?," paper delivered at the First Abraham Kuyper Consultation on "Common Grace: Theology, Ecology, and Technology," Princeton Theological Seminary, Princeton, NJ, February 2, 2002, pp. 11, 12.

[31] Roger Bate, *Pick a Number: A Critique of Contingent Valuation Methodology and its Application in Public Policy* (Washington, D.C.: Competitive Enterprise Institute, February 1994); William H. Desvousges, et. al. "Contingent Valuation: The Wrong Tool to Measure Passive-Use Losses," *Choices* (Second Quarter, 1993): 9-11; Peter A. Diamond and Jerry A. Hausman, "On Contingent Valuation Measurement of Nonuse Values," in Jerry A Hausman, ed. *Contingent Valuation: A Critical Assessment* (New York: North Holland, 1993); Steven F. Edwards, "Rethinking Existence Values," 68 *Land Economics* (February 1992): 120-22; Raymond J. Kopp, "Why Existence Value Should Be Used in Cost-Benefit Analysis," 11 *Journal of Policy Analysis and Management* (Winter 1992): 123-30; John Quiggin, "Existence Value and Benefit-Cost Analysis: A Third View," 12 *Journal of Policy Analysis and Management* (Winter 1993): 195-199; Alan Randall, "Passive-Use and Contingent Valuation C Valid for Damage Assessment," *Choices* (Second Quarter, 1993): 12-15; Donald H. Rosenthal and Robert H. Nelson, "Why Existence Values Should Not Be Used in Cost-Benefit Analysis," 11 *Journal of Policy Analysis and Management* 11 (Winter 1992): 116-22; Richard Stewart, *Natural Resource Damages: A Legal, Economic, and Policy Analysis* (Washington, D.C.: National Legal Center for the Public Interest,September 1995).

[32] Anonymous, "'Ask a Silly Question,...'= Contingent Valuation of Natural Resource Damages," 105 *Harvard Law Review* (June 1992): 1981-2000; Charles J. DiBona, "Assessing Environmental Damage," *Issues in Science and Technology* (Fall 1992).

[33] Jerry A. Hausman, ed., *Contingent Valuation: A Critical Assessment* (New York: North Holland, 1993).

[34] Richard T. Carson, et. al., *A Contingent Valuation Study of Lost Passive Use Values Resulting from the Exxon Valdez Oil Spill*, Report to the Attorney General of Alaska, Natural Resource Damage Assessment, Inc. (La Jolla, California, November 1992).

[35] Kenneth Arrow, Robert Solow, Paul R. Portney, Edward Leamer, Roy Radner, and Howard Schuman, *Report of the NOAA Panel on Contingent Valuation*, 58 Federal Register 4601 (January 15, 1993).

[36] Paul R. Portney, "The Contingent Valuation Debate: Why Economists Should Care," 8 *Journal of Economic Perspectives* (Fall 1994): 3-17.

[37] Jacoby, What is Nature Worth?," p. 20.

[38] Paul Milgrom, "Is Sympathy an Economic Value?: Philosophy, Economics, and the Contingent Valuation Method," in Hausman, ed., *Contingent Valuation*.

[39] John McPhee, *Encounters with the Archdruid: Narratives about a Conservationist and Three of his Natural Enemies* (New York: Farrar, Straus, and Giroux, 1971), p. 84.

[40] Roderick Nash, *Wilderness and the American Mind* (New Haven: Yale University Press, 1973), pp. 126.

[41] Quoted in Nash, *Wilderness and the American Mind*, p. 125.

[42] Quoted in Nash, *Wilderness and the American Mind*, p. 127.

[43] Roger G. Kennedy, "The Fish That Will Not Take Our Hooks," *Wilderness* (Spring 1995), p. 28.

[44] Mark Sagoff, "On the Expansion of Economic Theory: A Rejoinder," *Economy and Environment*, (Summer 1994), pp. 7-8.

[45] See Kate Soper, *What is Nature?* (Cambridge, MA: Blackwell, 1995).

[46] See Robert H. Nelson, *Public Lands and Private Rights: The Failure of Scientific Management* (Lanham, MD: Rowman and Littlefield, 1995); Robert H. Nelson, *A Burning Issue: A Case for Abolishing the U.S. Forest Service* (Lanham, MD: Rowman and Littlefield, 2000).

[47] See Robert Elliot, *Faking Nature: The Ethics of Environmental Restoration* (New York: Routledge, 1997).

[48] "The Exhibition of 1851," The The Speech of H.R.H. The Prince Albert, K.G., F.R.S., at The Lord Mayor's Banquet, in the City of London, October 1849. Printed in *The Illustrated London News,* 11 October 1849. [1849-10-11].

[49] Alejandro R. Roces, "Disasters are Getting Worse," *The Philippine Star* (Manila, the Philippines), July 3, 2007, p. 11.

[50] See Robert H. Nelson, *Two Gods: Economic Religion versus Environmental Religion, The New Terrain of Christian Theology* (forthcoming, 2009).

[51] On the powerful implicit normative elements in "scientific" economics, see Donald N McCloskey, "The Rhetoric of Economics," *Journal of Economic Literature* (June 1983).

[52] Frank R. Baumgartner, "Punctuated Equilibrium Theory and Environ-

mental Policy," in Robert Repetto, ed., *Punctuated Equilibrium and the Dynamics of U.S. Environmental Policy* (New Haven, CN: Yale University Press, 2006), p. 42.

[53] Helen Ingram and Leah Fraser, Path Dependency and Adroit Innovation: the Case of California Water," in Repetto, ed., *Punctuated Equilibrium and the Dynamics of U.S. Environmental Policy*, p. 81.

SOME THEOLOGICAL PERSPECTIVES ON THE CLIMATE CHANGE DEBATE

E. Calvin Beisner

I have been asked to speak to you on theological perspectives on the climate change debate. I shall do so, unapologetically. I speak intentionally and explicitly as a Christian committed to the Bible as the Word of God and as the source of the world view that enables us properly to understand God's creation. I shall speak on the basis of my Biblical presuppositions, fully aware that many of you do not share them. You think I am wrong about them, and I think you are wrong about them. So be it. This is not the context in which to debate that, but just as I do not protest your speaking on the basis of your unbiblical or antibiblical presuppositions, I trust that you will not protest my speaking on the basis of my Biblical presuppositions. There is no neutrality. All of us interpret all of our experiences, including scientific data, in light of our presuppositions. Failure to recognize that is not objectivity but philosophical naïveté.

As national spokesman for the Cornwall Alliance for the Stewardship of Creation, I facilitate discussion of the interrelated issues of population, resources, environment, and economic development. The Cornwall Alliance seeks to promote the application of Biblical theology, world view, and ethics, with sound science and sound economics, to the development of policies and practices to promote simultaneously economic development for the world's poor, especially in sub-Saharan Africa, and stewardship of creation. We believe poverty is the greatest threat to environmental quality and that sound policy therefore will seek to lift societies out of poverty so that they can go through the environmental transition we have already enjoyed: rising pollution in early development offset by improved health, life, and other standard measures of human well-being, followed by declining pollution in later stages of economic growth as people become willing and able to allocate more of their wealth to solving environmental problems. To put it simply, a clean, healthful, beautiful environment is a costly good and, like any other costly good, it is more affordable to the rich than to the poor. One fairly specific implication of our thinking on this is that policies to reduce forecast future temperatures that result in slower economic growth, especially for the world's poorest countries, will be counterproductive, delaying both the overcoming of the direct problems of poverty, such as disease and early mortality, and the indirect problem of poverty that is environmental abuse. Because most if not all policies aimed at reducing future temperatures by reducing carbon dioxide emissions will indeed slow economic development to the extent that they actually are implemented, this means that we generally oppose such policies.

I want to make a few comments prompted by my listening to earlier presentations in this conference. First, we have seen the tip of the iceberg of a major discussion that needs to be pursued in depth, though time does not permit it now: *the differences in perspective between modelers and empiricists*. Repeatedly in the history especially of environmental issues, the two approaches to science have come down on opposite sides of major questions. One example other than global warming is species extinction rates. Modelers claim high and rising rates, but field observations, even those commissioned by the International Union for the Conservation of Nature in the effort to support such model-generated claims, fail to support them.[1] As one who teaches historical theology, I find it ironic to see in this divide an analogy to the struggle during the development of medieval science to escape preconceived ideas. Interestingly enough, today's modelers seem to be operating in the tradition of the medieval theologians and philosophers who insisted that human anatomy and other matters really did not need study by direct observations because Aristotle and Galen already had described them quite sufficiently.

Second, as an interdisciplinary scholar—one trained and committed to the symphonic application of the principles, methods, and tools of several disciplines to address multifaceted problems—I sense that the more narrowly focused those studying this problem are, the more likely they are to embrace the more alarming scenarios and the more drastic proposed solutions. Those who look at wider arrays of evidence from assorted disciplines seem generally to take more moderate views and suggest more moderate solutions. I think that there is considerable need for more interdisciplinary study—not just study of different aspects of the large problem by different scholars, but an attempt by individual scholars to study many aspects of the problem. A good example of that is the symphonic array of evidences brought to bear on long-term climate cycles by Fred Singer and Dennis Avery in their 2006 book, *Unstoppable Global Warming—Every 1,500 Years*.

Third, as a professor of logic I often find myself noting logical fallacies in discussions of these matters. Positing unforeseen disasters as rationales for policy, for example, commits the twin fallacies of *argumentum ad silencio* and *argumentum ad futuram*. The appeal to consensus, even of experts when well-qualified experts disagree, is a fallacy of *consensus gentium*. The appeal to experts not for evidence and explanation in their own field but for conclusions in another is *argumentum ad verecundiam* (illegitimate appeal to authority). Attending only to evidence that seems to confirm a hypothesis but ignoring contrary evidence—a practice that seems inextricably embedded in the very mandate of the Intergovernmental Panel on Climate Change (IPCC)—is an inductive fallacy.

Fourth, in light of three speakers' comments that population growth is a serious problem—and note, by the way, that none of them was a demographer or an economist of demography—let me also point out that the greater problem related to population appears now to be the prospect of its shrinking. Aside from the fact that professional demographers have essentially abandoned all hope of obtaining an objective definition of "overpopulation" (neither density nor growth rate works), fertility rates are now below replacement level in all high-income and many middle-income countries and fell below replacement level in twenty developing countries by 2002. Europe is losing about 800,000 people per year and Russia about a million. Fertility rates are falling faster in the developing nations than they did during the same stages of development in the presently developed nations. Professional demographers, such as those at the U.N. Population Division (which is to be distinguished from the much more politicized U.N. Fund for Population Activities), estimate that world population will peak sometime between 2035 and 2065 (I lean toward the earlier time) at something between seven and eight billion (I lean toward the lower figure)—and then it will begin to shrink, increasingly rapidly. If the underlying causes of the fertility decline—rooted in changing fertility preferences related to economic development—do not change drastically, there is every possibility that, as Stanley Kurtz put it in *Policy Review* two-and-a-half years ago, "If worldwide fertility rates reach levels now common in the developing world (and that is where they seem headed), within a few centuries, the world's population could shrink below the level of America's today."[2] And should someone want to suggest that sub-Saharan Africa is overpopulated, I might point out that its population density is far below the global average, far below that of any European nation, and far below America's. Its population problem, to the extent that it has one, is population density too low to create enough tax base to fund infrastructure necessary for greater economic development.

Theological and Scientific Aspects of Al Gore's Work

To close out these preliminary remarks, in light of Al Gore's having received both an Oscar and a Nobel Peace Prize for his film *An Inconvenient Truth*, let me comment just briefly and directly on the religious and political aspects of his work. It is clear from his film and, much more importantly, from his 1992 book *Earth in the Balance*, that he approaches global warming (and other environmental issues) as matters of religious concern. But it is a mistake to think of his religion as Christian. I read *Earth in the Balance* very carefully when it first came out, and while it is loaded with religion, it is not Christian religion. It is pagan, New Age, Gaia religion through and through. If "Christian" means "nice" or "American" or "caring," then perhaps Gore is a Christian; but if the word has any doctrinal content, any reference to the great ecumenical creeds of the early Church or to the confessions of the Reformation era, he is not a Christian in that sense.

Further, Gore is a political totalitarian. In *Earth in the Balance*, he wrote, "I propose a program involving as many countries as possible that will use schoolteachers and their students to monitor the entire earth daily."[3] The proposal is one tiny part of his sweeping "Global Marshall Plan" to stave off environmental disasters. It seems insignificant compared with others. But it reveals both Gore's political instincts and his inability to see the consequences of his ideas. Never mind the absurdity of further burdening schools that fail to teach children basic skills. More ominous is Gore's failure to realize that he has proposed that all students be drafted into involuntary servitude. Sadly, most of the rest of Gore's proposals are consistent with these statist political instincts. The "Global Marshall Plan" consists of five "strategic goals" (summarized in *Earth in the Balance*, pp. 305-307) and the means of achieving them:

1. Stabilizing world population ("No goal is more crucial . . .") by raising literacy rates, lowering infant mortality, and increasing access to birth control techniques;
2. "Rapid creation and development of environmentally appropriate technologies" by subsidizing research and development and the transfer of technologies to developing countries (Who pays for this transfer?);

3. Comprehensive changes in economic measurements (e.g., to reflect environmental damage and resource depletion as negatives in measures of GDP or GNP—a sensible idea in principle but frequently applied senselessly);

4. "Negotiation and approval of a new generation of international agreements that will embody regulatory frameworks, specific prohibitions, enforcement mechanisms, cooperative planning, sharing arrangements, incentives, penalties, and mutual obligations necessary to make the overall plan a success" (The implications for increased taxes, bureaucracies, regulations, and threats to national sovereignty are mind-boggling);

5. "Establishment of a cooperative plan for educating the world's citizens about our global environment." Shades of *1984* arise. What will happen to dissenting voices when global intergovernmental agreements promote a single view of the alleged crises? When I first asked that question shortly after the book came out, I was largely speculating. Now, of course, we have had calls for "Nuremberg trials" for "climate deniers" and for stripping meteorologists who reject catastrophic anthropogenic global warming of their credentials, and the actual stripping of state climatologists of their offices if they reject it.

Fifteen years ago, as now, Gore stared inconsistencies straight in the eye and ignored them. On p. 301, he wrote that the "framework of global agreements" must "obligate all nations to act in concert . . . while carefully respecting the integrity of individual nation-states." But on p. 302 he wrote that the agreements "will contain both incentives and legally valid penalties for noncompliance."

His inconsistency didn't stop at political theory. He had (and still has) a tough time with numbers, too. The text on p. 24 says that we are extinguishing species "a thousand times faster than at any time in the past 65 million years (see illustration)"—and the illustration suggests a past rate of around one or two per year, implying a current rate of 1,000 to 2,000 per year. But the illustration itself suggests a current rate of 8,000 to 10,000 per year (5 to 10 times that claimed by the text). The text on p. 28 says 100 per day, *i.e.*, 36,500 per year (36.5 to 18.25 times that claimed by the text on p. 24). And the text on p. 120 implies (apparently several) thousand per year. (Reliable field studies indicate a much lower rate.) Gore did not present these different rates as alternative possibilities; he presented them all as if they were true. Extending his graph, by the way, entails predicting all species extinct by about 2015.

Gore's trouble with numbers involves more than inconsistency. Explicitly in public speaking and implicitly in the book (p. 38), Gore claimed that virtually all of the scientific community ("98 percent") shared his view of the reality, magnitude, and dangers of global warming ("the most serious threat that we have ever faced" [p. 40]). Yet a fall 1991 Gallup survey of 400 climatologists, atmospheric scientists, and oceanographers showed that only 41 percent believed the theory was substantiated by current evidence, 31 percent did not accept it, and 28 percent said they did not know. No matter. Those who disputed Gore's views were merely "self-interested cynics . . . seeking to cloud the underlying issue . . . with disinformation" (p. 360), not "reputable scientists" (p. 89).

If scientists do not agree with you, you can make it appear that they do. Gore wrote that Richard Lindzen, a critic of Gore's view of catastrophic anthropogenic global warming, "publicly withdrew his hypothesis on how [increasing evaporation might counteract global warming] in 1991" (p. 90). Only if you had the patience to read his hopelessly uninformative notes did you discover that Lindzen still held his basic theory (p. 380); he had only withdrawn one hypothesis about how it worked. Talk about disinformation!

And if you cannot make it appear that the scientists agree, you can silence them. According to Candace C. Crandall of the Science and Environmental Policy Project, "Gore [had] been quietly suggesting to journalists that they suppress environmental evidence that is not alarming. Publicizing such evidence would, in Gore's words, 'undermine the effort to build a solid base of public support for the difficult actions we must soon take.'"[4]

Those things said by way of introduction, let me move to the body of my address, and generally to more theological matters. But please do not be shocked if I also discuss science and cite some scientific sources; theologians can read science, too, after all.

Creation, Creator, and Doxology

On August 9, 2007, Cornwall Alliance contributing scientist and University of Alabama senior research climatologist Roy W. Spencer, with three coauthors, published in the *Journal of Geophysical Research* an article with the typically opaque scientific title "Cloud and radiation budget changes associated with tropical intraseasonal oscillations."[5] That intimidating title masked the fascinating and momentous finding that high-altitude cirrus clouds in the tropics diminish rather than increase with rising surface temperatures and thus are not a positive (reinforcing) but a negative (reducing) feedback on climate change. That is, their response to rising temperature is somewhat like the response of the iris of the eye to light. Spencer et al.'s research, the conclusion of which is directly opposite to the assumptions used in all computer models of climate change, thus lends support to a theory by meteorologist Richard Lindzen that the atmosphere acts like an iris to regulate surface temperatures.[6]

Although both findings have stunning implications for the ongoing debate about global warming—because they show that models projecting major warming in response to increased carbon dioxide assume that this feedback does precisely the opposite of what it actually does—those implications are not why I cite them here. Rather, I cite them for their doxological effect. They ought to prompt us to praise the Creator for His great wisdom in creation.

The Bible begins, "In the beginning God created the heavens and the earth." When we seek understanding of any subject, we can do no better than to start with God, who reveals Himself in Scripture and creation. "The heavens are telling of the glory of God; and their expanse is declaring the work of His hands. Day to day pours forth speech, and night to night reveals knowledge" (Psalm 19:1-2). Praise for the Creator should be our first and highest response to His creation.

Just what, though, does creation itself reveal to us about the Creator? His greatness, His glory, surely. But greatness and glory in what? In all of His attributes, no doubt, but for today let us focus on just two: His wisdom and power.

God's wisdom and power shine in the complexity of design, from the *invisible* micro level of the irreducibly complex molecular machines of proteins described in biochemist Michael Behe's *Darwin's Black Box: The Biochemical Challenge to Evolution*; to the *visible* middle level of our everyday lives in which we observe complex social and ecological interactions among people, plants, and animals; to the *invisible* macro level's literally and incomprehensibly complex fluid dynamics of the ocean/atmosphere climate system described in *Taken By Storm: The Troubled Science, Policy and Politics of Global Warming*, by applied mathematician Christopher Essex and environmental economist Ross McKitrick (a Cornwall contributing scholar), both of whom have specialized in mathematical and statistical analysis of climate.[7]

Reading *Taken By Storm*'s brilliant discussion of climate as a turbulent fluid system in light of the mathematics of mysterious, in principle unsolvable differential equations, the even more mysterious Navier-Stokes equations, and the puzzling "Kolmogorov's trash can," all of which help us to recognize that there is as much "chaos"—and consequently as much unpredictability—at the macro level of climate as there is at the micro level of quantum physics, should move any Christian or Jew to praise the Creator of this amazing system. Essex and McKitrick explained that "[t]o do the fluid dynamics correctly . . . requires, in addition to the basic ([mathematically] impossible) turbulence problem, tracking [chaotically moving] filaments in detail within the flow." But both of those are impossible not because we lack computing or observing power but because many of the equations are simply unsolvable in principle. All of this entailed that even if we had quadrillions of tiny temperature sensors evenly distributed through all the atmosphere and oceans, with each reporting its temperature every millisecond to a computer of infinite capacity and speed, it would still be impossible for that computer to make credible projections of future climate.

Their summation—"We can't begin to do this . . ." (p. 81)—should remind us of Job 38 and move us to praise the wisdom and power of the God who designed things that not only exceed our present knowledge but also, by their very nature, like the subatomic particles of quantum physics, cannot possibly be known by finite minds, yet are entirely under His sovereign control.[8]

As we think about creation stewardship, then, the very first thing we must keep in mind is the doctrine of God—particularly, that an infinitely wise, infinitely powerful Creator made and sustains the universe and every part of it. From this we can infer that the design of all things reflects the wisdom of God, and the sustaining of all things reflects the power of God. These truths are immediately relevant to creation stewardship in general and to the climate change debate in particular.

Much environmental writing assumes that the Earth and its various ecosystems are extremely fragile, subject to immense and irreversible damage from even tiny changes. Ironically, many environmentalists derive the notion of Earth's fragility from the first of Barry Commoner's "Four Laws of Ecology": "Everything is connected to everything else."[9] Therefore, they reason, a little disruption here is bound to cause disruption everywhere. But the opposite inference can be made: Because everything is connected to everything else, even major disruption here is compensated for by minor adjustments everywhere.[10]

Which of these inferences from "Everything is connected to everything else" is more consistent with the Bible's teaching that God is a wise Creator? The common environmentalist view of a fragile planet, or the view that, though the earth is vulnerable to harm, its consummate design makes it resistant and resilient, thus minimizing the risk of catastrophic and irreversible damage? Which of these inferences seems more consistent with the finding by Spencer and his co-authors that tropical cirrus clouds are a negative rather than a positive feedback on surface temperatures? Is it more likely, in light of the Bible's teaching about creation, that a perturbation of Earth's climate system will set off a "runaway greenhouse effect," or that it will be compensated for by myriads of tiny adjustments throughout the atmosphere, hydrosphere, cryosphere, biosphere, and even the geosphere, all acting together as a self-regulating system? Does a wise engineer design a system so that positive feedbacks to a new influence will vastly outweigh negative feedbacks—the assumption behind projections of high gains in temperature from a doubling of atmospheric CO_2—thus inviting the whole system to career off into irretrievable chaos with the slightest change? Or does he design a system so that positive and negative feedbacks tend to maintain reasonable stability—the assumption behind projections of moderate warming from enhanced CO_2? Which of those understandings of the earth and its systems is more consistent with God's promise to Himself in Genesis 8:22, "While the earth remains, seedtime and harvest, and cold and heat, and summer and winter, and day and night will not cease?" Or with His promise in the covenant with Noah after the Flood that "all flesh shall never again be cut off by the water of the flood, neither shall there again be a flood to destroy the earth" (Genesis 9:11)? Or with the Psalmist's statement that God "established the earth upon its foundations, so that it will not totter forever and ever" and that following the Flood He "set a boundary that [the waters] may not pass over, so that they will not return to cover the earth" (Psalm 104:5, 9)?

167

What about Global Warming?

Yet global warming, with sea level rise as its most disastrous consequence, is clearly the one environmental problem that dominates thinking lately, especially with Al Gore and the IPCC sharing this year's Nobel Peace Prize (announced, ironically, just a week after a British high court judge found that Gore's film, *An Inconvenient Truth*, has nine major scientific errors and therefore may be shown in British schools only if accompanied by literature correcting those errors). The global warming hypothesis is that human emissions of greenhouse gases are causing unnatural increases in global temperature and thus threatening catastrophic impacts in the form of rising sea level, more and stronger hurricanes and other severe weather events, droughts, floods, crop failures, species extinctions, and the spread of tropical diseases. The issue deserves careful consideration, particularly because it is touted by some, like Gore, as the greatest threat ever to face humanity.

The popular conception is that this hypothesis has achieved the status of overwhelming scientific consensus and is supported by overwhelming scientific evidence. Is either of these the case? The fact that I approach these things as a theologian does not mean that I should ignore science, after all.

The Myth of Scientific Consensus

As we saw above, Al Gore claimed overwhelming scientific consensus for carbon-accelerated global warming in the early 1990s—despite the contrary evidence of an actual poll of climate scientists. The same claim continues today. Is it true?

The best evidence for the claim is a paper by Naomi Oreskes, et al., published in *Science* in 2004, that claimed to have found that no papers on the ISI Web of Science database of refereed publications from 1993 through 2003 rejected what she called the consensus that "most of the observed warming over the last 50 years is likely to have been due to the increase in greenhouse gas concentrations" (IPCC, 2001).[11] Yet another scholar, Benny Peiser, attempting to replicate the results from the same database, discovered serious flaws in Oreskes's method and concluded that no such consensus existed in the refereed literature. Instead, he found that:

- Only 1 percent explicitly endorsed what Oreskes called the "consensus view";

- 29 percent implicitly accepted it "but mainly focus[ed] on impact assessments of envisaged global climate change";
- 3 percent rejected or doubted it; and
- 42 percent did "not include any direct or indirect link or reference to human activities, CO_2 or greenhouse gas emissions, let alone anthropogenic forcing of recent climate change."[12]

Controversy has raged between advocates and critics of the manmade catastrophic warming hypothesis over whether Oreskes's or Peiser's findings are more trustworthy. In my judgment, the patent methodological flaws Peiser pointed out in Oreskes's study vitiate her results. Peiser's findings also are more consistent with the results of a 2003 survey of climate scientists. They were asked, "To what extent do you agree or disagree that climate change is mostly the result of anthropogenic causes?" Of the 530 valid responses, 9.4 percent strongly agreed, while 9.7 percent strongly disagreed. These results and the mean of 3.62 (out of 7) demonstrate that among climatologists consensus is not strong that climate change is mostly the result of anthropogenic causes.[13]

A new study by Klaus-Martin Schulte of the same database used by Oreskes and Peiser, this time covering the period 2004 through early 2007, found (assuming that Oreskes's findings were correct) that the proportion of papers endorsing what she had called the consensus had fallen from 75 to 45 percent, while the proportion rejecting it had risen from 0 to 6 percent, and 42 percent are uncommitted. "The results," says Schulte, "show . . . a significant movement of scientific opinion *away from* the apparently unanimous consensus which Oreskes had found in the learned journals from 1993 to 2003."[14] Additional evidence against consensus appears in D. Bray and H. Von Storch's recently published *The Perspectives of Climate Scientists on Global Climate Change*, which concludes, "As the data seems to suggest, the matter is far from being settled in the scientific arena."[15] In short, scientific consensus on catastrophic anthropogenic global warming (as distinct from climate change generally, by whatever causes) is a myth.

Further, as my coauthors and I pointed out in "A Call to Truth, Prudence, and Protection of the Poor: An Evangelical Response to Global Warming," unlike politics, but like truth, science is not a matter of consensus but of data and valid arguments. As Thomas Kuhn so famously pointed out in *The Structure of Scientific Revolutions*, great advances in science, often involving

169

major paradigm shifts, occur when small minorities patiently—and often in the face of withering opposition—point out anomalies in the data and inadequacies in the reigning explanatory paradigms until their number and weight become so large as to require a wholesale paradigm shift, and what once was a minority view becomes a new majority view. Indeed, skepticism is essential to science. As Robert Merton put it almost 70 years ago, "Most institutions demand unqualified faith; but the institution of science makes skepticism a virtue."[16]

Conflicting Scientific Evidence

Why has support for the alleged "consensus" view declined, as Schulte's findings indicate? Primarily, I think, because of new studies either reducing the apparent role of human contribution or magnifying the apparent role of non-human contributors to climate change. The Cornwall Alliance has published several papers citing such findings.[17] Consider nine representative samples of such studies and their findings:

1. The 2007 *Scientific Assessment* of the IPCC reduced its estimate of global temperature response to doubled CO_2 by 25 percent, its estimate of human contribution to energy absorption in the atmosphere by 35 percent, and its projection of sea level rise by as much as 50 percent, versus its 2001 *Assessment*'s forecasts.
2. The Sea Level Commission of the International Union for Quaternary Research reports that the most likely sea level rise over this century is only 0 to 7.88 inches, not the 17 to 23 inches claimed by the *Fourth Assessment Report*.
3. In June 2007, a study by atmospheric scientist Stephen Schwartz of Brookhaven National Laboratory concluded that doubled CO_2 would only raise global average temperature by about 1.1° C, not the roughly 3° C assumed by the IPCC.[18]
4. Increasing evidence of many sorts pointed to several overlapping cycles of global warming and cooling from entirely natural causes that overshadow anthropogenic warming and explain the warming of the late 20th century.[19]
5. The publication in early 2007 of *The Chilling Stars: A New Theory of Climate Change*, by Danish solar physicist Henrik Svensmark and Nigel Calder,[20] provided plausible evidence that much recent and

170

longer-term global temperature change is explained by fluctuations in solar energy and solar magnetic wind output and the latter's interaction with cosmic ray flux and thus the formation of cloud nuclei—although the precise physical explanation of how this works remains difficult, the correlations are strong enough to make coincidence an uninviting hypothesis.

6. The ubiquitous claim that 1998 was the hottest year on record and the 1990s were the hottest decade on record for U.S. surface temperatures was debunked. Instead, it turned out that 1934 was the hottest year and the 1930s were the hottest decade. The ten hottest years since 1880 are now, in descending order, 1934, 1998, 1921, 2006, 1931, 1999, 1953, 1990, 1938, 1939, with three of the top ten in the last decade but four in the 1930s.[21] The consequence? First, that at least for the U.S., there is no significant upward temperature trend since the 1930s. Second, because the U.S. temperature data had generally been considered the most comprehensive and reliable, the finding that they had been thrown off by a programming error calls into question temperature data worldwide.[22]

7. Spencer and coauthors, as we saw above, found that tropical cirrus clouds are a negative rather than positive feedback on surface warming. How significant was their finding? Says Spencer, "To give an idea of how strong this enhanced cooling mechanism is, if it was operating on global warming, it would reduce estimates of future warming by over 75 percent."[23]

8. It was revealed that a paper on which the IPCC relied heavily for its 2007 *Fourth Assessment Report*'s estimate of surface temperature change was "based on fabricated data." The paper[24] "is one of the main works cited by the IPCC to support its contention that measurement errors arising from urbanization are tiny, and therefore are not a serious problem." It in turn relied on another paper by one of its own authors.[25] Both papers claimed that they carefully used data only from meteorological stations "with few, if any, changes in instrumentation, location or observation times"—important because such changes make data incomparable over time. Those two papers in turn cited as their source a report resulting from a project done jointly by the U.S. Department of Energy (DOE) and the Chinese Academy of Sciences.[26] But that report explicitly said that station histories were not available

171

for 49 (58 percent) out of the 84 Chinese meteorological stations used. "For those 49 stations, then, the above-quoted statements from the two papers are impossible," points out Douglas J. Keenan, who goes on to point out serious discontinuities in the remaining 35 stations as well. Keenan concludes: "The essential point here is that the quoted statements from Jones et al. and Wang et al. cannot be true and could not be in error by accident. . . . The statements are fabricated," adding, "The conclusions are clear. First, there has been a marked lack of integrity in some important work on global warming that is relied upon by the IPCC. Second, the insignificance of urbanization effects on temperature measurements has not been established as reliably as the IPCC assessment report assumes." (Keep this in mind the next time you hear of the IPCC as a peer-reviewed process—and indeed the next time you think peer review ensures accuracy.[27])

9. Tsonis et al. proposed a whole new theory to explain climate shifts. As synchronized chaos theory in mathematics explains, a periodic synchronization of known Earth ocean cycles (Pacific Decadal Oscillation, North Atlantic Oscillation, El Niño, and North Pacific Oscillation) could explain the major climate shifts observed thus far without reference to any trends in greenhouse gases.[28]

The point is not that any one of these other influences on climate alone explains all changes but that as our knowledge of alternative influences grows, we need to reduce our estimates of the role of anthropogenic influences. In short, these developments provide good reason to reject or at least to question seriously the popular claim that human action is driving catastrophic climate change and to conclude instead, as the Cornwall Alliance's "Call to Truth" did, that recent and foreseeable climate change are cyclical, largely (even if not totally) natural, well within the bounds of historic variability, and neither already nor likely to become catastrophic.

Prioritizing Problems: Learning from the Copenhagen Consensus

The Bible puts considerable emphasis on protecting the poor and meeting their needs (e.g., Galatians 2:10). One condition of doing that is understanding what things cause the greatest harm to the poor and focusing our efforts on those rather than on other risks that are less important to them.

The Copenhagen Consensus Center of the Copenhagen Business School commissioned papers by world-class experts on challenges facing human- ity. For each such paper, it commissioned responses from two additional experts who disagreed in significant ways with the primary author. Finally, it submitted the resulting papers to a panel of eight world-class economists,[29] including four Nobel laureates, to compare the risks associated with the problems and the benefits that could be expected from various means of ad- dressing them. This panel then ranked the proposals described in the papers, listing them under four categories: very good, good, fair, and bad—meaning projects in which costs exceed benefits.

Of the top 17 opportunities, the three worst (all categorized as bad be- cause costs outweighed benefits) all had to do with attempting to reduce global warming.[30] The greatest opportunities for achieving real results in environmental spending appear to be in addressing communicable diseases (particularly AIDS and malaria) and sanitation and water (particularly community-based drinking water supply and purification, small-scale wa- ter technology for livelihoods, and research on water productivity in food production). Millions of poor people in developing countries die every year because they lack clean water and indoor plumbing (exposing them to water- borne diseases like dysentery and other diarrheal diseases that cause some 1.8 million premature deaths per year); because they lack electricity (forcing them to burn wood and dung for cooking and heating and to live without refrigeration and air conditioning and thus to breathe indoor air pollution that causes some 1.6 million premature deaths and scores of millions of serious respiratory illnesses annually and to either eat spoiled food and risk illness or throw it away and risk hunger);[31] and because they lack sewage treatment, jobs, access to affordable medical care, and adequate nutrition.

To put it briefly and simply: the greatest threat to the environment is poverty. Poverty drives high per-capita and per-unit-of-production pollution emission rates, low pollution-cleanup rates, and the high rates of disease and death associated with all of these, not to mention waste of resources, deforestation, and habitat loss. The implication is clear: economic develop- ment is the most important environmental task facing us today. In light of the graphics presented yesterday showing that lower latitudes, especially in sub-Saharan Africa, are likely to be the most hurt by warming, and in light of the fact that wealth makes adaptation to any climate easier, it seems clear that we must strive to promote economic development, especially in such places.

Conclusion

I close by sketching briefly one other theological insight relevant to the anthropogenic global warming debate. Biblical thought recognizes that God sometimes teaches us through typology, in which one thing, through analogy, functions as a type of, or pointer to, another. Some Biblical types point toward the Messiah, in whom they are fulfilled. Adam as representative of all his descendants is typical of the Messiah as representative of all for whom He died sacrificially. Joseph, betrayed by his brothers and yet exalted to honor by God, is representative of the Messiah similarly betrayed and exalted. The sacrificial lamb in the priestly system is representative of the Messiah who came as the "Lamb of God, who takes away the sins of the world." Early Christians also pointed to types of Christ in the world around us—e.g., Tertullian saw figures of the cross in the shape of a bird in flight, the mast of a ship, and many other familiar shapes.

I think there is a wonderful type in the carbon cycle. Where does coal, the source of a great deal of our carbon emissions, come from? Great masses of plants and animals were buried and through heat and pressure turned to coal. Now we dig up the coal, burn it, and make from it energy that serves our needs plus carbon dioxide that enhances plant growth—every doubling of dioxide causing about a 35 percent increase in plant growth efficiency, resulting in higher crop yields and better survival for all kinds of plants in all climatic conditions. Any Christian should recognize the figure: death, burial, and resurrection to impart new life to others. That is what we see in Jesus Christ. It is sad indeed that many, instead of recognizing that type and rejoicing in the life-giving effects of enhanced carbon dioxide, respond only in fear of its heat-trapping effect. The typology doesn't prove the case, but it provides a lovely mental image for what the science tells us already: that in the carbon cycle we get life from death.

While I recognize that plenty of people see things differently, a good case can be made that a Biblical world view, consistently applied to a wide variety of scientific evidences, will lead to the general conclusion that recent and foreseeable climate change are largely natural in cause, cyclical, well within the bounds of historic variability, and neither now, nor likely to become, catastrophic. Also, rather than trying to mitigate global warming, we shall do much more good for humanity and the rest of the environment by promoting human economic development, thus enabling us to adapt to any climate future, whether warmer or cooler. Meanwhile, we should address directly the many problems for people and the environment created by poverty.

Endnotes

[1] Julian L. Simon and Aaron Wildavsky, "Disappearing Species and the Absence of Data," in Julian L. Simon and Herman Kahn, eds., *The Resourceful Earth: A Response to Global 2000*, Oxford, UK, and Cambridge, MA: Blackwell (1984); T.C. Whitmore and J.A. Sayer, eds., *Tropical Deforestation and Species Extinction*, London, UK: Chapman and Hall (1992); and Julian L. Simon and Aaron Wildavsky, "Species Loss Revisited," in Julian L. Simon, ed., *The State of Humanity*, Oxford, UK, and Cambridge, MA: Blackwell (1995).

[2] Stanley Kurtz, "Demographics and the Culture War," *Policy Review*, February/March 2005, online at http://www.hoover.org/publications/policyreview/3431156.html. *See also*, Ben Wattenberg, *Fewer: How the New Demography of Depopulation Will Shape Our Future*, Chicago, IL: Ivan R. Dee (2005).

[3] Albert Gore, Jr., *Earth in the Balance: Ecology and the Human Spirit*, New York, NY: Houghton Mifflin (1992), p. 356.

[4] Candace C. Crandall, "His finger is on the panic button," Orange County Register, August 9, 1992, Section J, pp. 1-2.

[5] R.W. Spencer, W.D. Braswell, J.R. Christy, and J. Hnilo (2007), "Cloud and radiation budget changes associated with tropical intraseasonal oscillations," *Geophysical Research Letters*, vol. 34 (2007), L15707, doi: 10.1029/2007GL029698, abstract online at http://www.agu.org/pubs/crossref/2007/2007GL029698.shtml.

[6] Richard S. Lindzen, Ming-Dah Chou, and Arthur H. Hou, "Does the Earth Have an Adaptive Infrared Iris?" *Bulletin of the American Meteorological Society*, vol. 82, no. 3 (March 2001), pp. 417-432.

[7] Christopher Essex and Ross McKitrick, *Taken by Storm: The Troubled Science, Policy, and Politics of Global Warming*, Toronto, Ont.: Key Porter Books (2002), chapter 3, "Climate Theory versus Models and Metaphors."

[8] See R.C. Sproul, *Not a Chance: The Myth of Chance in Modern Science and Cosmology*, Grand Rapids, MI: Baker Academic (1999).

[9] Barry Commoner, *The Closing Circle: Nature, Man, and Technology*, New York, NY: Alfred A. Knopf (1971). The key passage on the "Four

Laws of Ecology" is online at http://www3.niu.edu/~td0raf1/history261/nov1910.htm.

[10] See Gregg Easterbrook, *A Moment on the Earth: The Coming Age of Environmental Optimism*, New York, NY: Viking (1995), chapters 4, 6, 8, and 10.

[11] Naomi Oreskes, "The scientific consensus on climate change," *Science*, vol. 306, no. 5702 (December 3, 2004), p. 1686, online September 21, 2007, at http://www.sciencemag.org/cgi/content/full/306/5702/1686.

[12] Benny J. Peiser, Letter to *Science*, January 4, 2005, submission ID: 56001, online September 21, 2007, at www.staff.livjm.ac.uk/spsbpeis/Scienceletter.htm.

[13] Letter from Dennis Bray to *Science*, December 22, 2004, not accepted for publication. Bray is a climatologist with the GKSS Forschungszentrum Geestracht, Germany, and a vocal critic of "global warming skeptics." His letter was online October 14, 2006, at http://www.staff.livjm.ac.uk/spsbpeis/Scienceletter.htm.

[14] Klaus-Martin Schulte, "Scientific Consensus on Climate Change?" unpublished paper provided to me by Sonja Boehmer-Christiansen, editor of *Energy and Environment*, who plans to publish a revised version of the paper in that journal (e-mail, Boehmer-Christiansen to Beisner, September 21, 2007).

[15] D. Bray and H. von Storch, *The Perspectives of Climate Scientists on Global Climate Change*, GKSS-Forschunszentrum Geesthacht GmbH (2007), p. 7.

[16] Robert K. Merton, "Science and the Social Order," *Philosophy of Science*, vol. 5, no. 3 (July 1938), pp. 321-337, at 334.

[17] E. Calvin Beisner, Paul K. Driessen, Ross McKitrick, and Roy W. Spencer, "A Call to Truth, Prudence, and Protection of the Poor: An Evangelical Response to Global Warming," Burke, VA: Interfaith Stewardship Alliance/Cornwall Alliance for the Stewardship of Creation (2006), online September 24, 2007, at http://www.cornwallalliance.org/docs/Call_to_Truth.pdf. This document has been endorsed by over 170 leaders, including many climatologists, meteorologists, and other topic-qualified scientists, and environmental and developmental economists, plus theologians, pastors, mission leaders, and other religious leaders. A list of endorsers may be seen at http://www.cornwallalliance.org/docs/Open_Letter.pdf. *See also* Beisner, "Important Developments on

Global Warming in 2006," online September 21, 2007, at www.interfai
thstewardship.org/pdf/GlobalWarmingSummary2006.pdf, and Beisner,
"Global Warming: Why Evangelicals Should Not Be Alarmed," *Reformed
Perspective*, vol. 21, no. 11 (September 2007), pp. 24-7; online at http:
//www.cornwallalliance.org/docs/Global-Warming–Why-evangelicals-
should-not-be-alarmed.pdf.

[18] Stephen E. Schwartz, "Heat Capacity, Time Constant, and Sensitivity of
Earth's Climate System," Brookhaven National Laboratory, June 2007,
online at www.ecd.bnl.gov/steve/pubs/HeatCapacity.pdf accepted for
publication in *Geophysical Research Letters*.

[19] S. Fred Singer and Dennis Avery, *Unstoppable Global Warming—Every
1,500 Years*, Lanham, MD: Rowman and Littlefield (2006).

[20] Henrik Svensmark and Nigel Calder, *The Chilling Stars: A New Theory
of Climate Change*, Cambridge, MA: Icon Books (2007).

[21] Raw data are at www.data.giss.nasa.gov/gistemp/graphs/Fig.D.txt.

[22] Readers who want to see extensive discussion of this and related problems
should visit the website of Canadian statistician Stephen McIntyre, who
discovered NASA's error (and was one of the discoverers of the errors
in the "hockey stick" graph mentioned earlier): www.climateaudit.org.

[23] See abstract of the *Geophysical Research Letters* article at http:
//www.agu.org/pubs/crossref/2007/2007GL029698.shtml and a
University of Alabama press release about it at www.uah.edu/news/
newsread.php?newsID=875, and www.blogs.usatoday.com/weather/
2007/08/cloudy-forecast.html.

[24] P.D. Jones, P.Y. Groisman, M. Coughlan, N. Plummer, W.-C. Wang, and
T.R. Karl, "Assessment of urbanization effects in time series of surface
air temperature over land," *Nature*, vol. 347 (1990), pp. 169-172.

[25] W.-C. Wang, Z. Zeng, and T.R. Karl, "Urban heat islands in China,"
Geophysical Research Letters, vol. 17 (1990), pp. 2377-2380.

[26] http://cdiac.esd.ornl.gov/ndps/ndp039.html.

[27] For the full story on this issue, see www.motls.blogspot.com/2007/08/
urban-heat-islands-fabricated-papers.html, http://www.informath.org/
WCWF07a.pdf, and http://www.informath.org/apprise/a5620.htm.

[28] For more, see Michael Asher, "Major New Theory Proposed to Explain
Global Warming," *DailyTech*, August 14, 2007, at www.dailytech.com/
Major+New+Theory+Proposed+to+Explain+Global+Warming/
article8450.htm, and Tsonis, *et al.*, "A new dynamical mechanism for ma-

jor climate shifts," *Geophysical Research Letters*, vol. 34 (2007), L13705, doi:10.1029/2007GL030288, at http://www.volny.cz/lumidek/tsonis-grl.pdf.

[29] Jagdish N. Bhagwati, Columbia University; Robert S. Fogel (Nobel laureate), University of Chicago; Bruno S. Frey, University of Zurich; Justin Yifu Lin, Peking University; Douglass C. North (Nobel laureate), Washington University in St. Louis; Thomas Schelling, University of Maryland (Nobel laureate); Vernon L. Smith (Nobel laureate), George Mason University; and Nancy Stokey, University of Chicago.

[30] Bjorn Lomborg, ed., *Global Crises*, p. 606 (for overall panel ranking), pp. 608-644 (for individual panelists' rankings).

[31] Alex Kirby, "Indoor smoke 'kills millions,'" BBC News, November 28, 2003; online September 24, 2007, at http://news.bbc.co.uk/2/hi/science/nature/3244214.stm.

ON GLOBAL WARMING HERESY*

Richard S. Lindzen

I am frequently asked to describe my experiences as a contrarian about global warming. I still find the request somewhat annoying, and in this piece I would like to explain why. For starters, to be a contrarian generally implies an automatic tendency to go against popular wisdom. That is not my position.

What in the world does it really mean to be a 'contrarian' on the issue of global warming? On an issue where virtually all popular depictions depend on long chains of uncertain connections, support for all these linkages would constitute more a religious faith than a scientific position. On the other hand, where the elements of the picture do deal with relatively basic issues, there is, in fact, little disagreement. Some examples may help clarify the situation.

For instance, there is little argument that levels of CO_2 in the atmosphere have risen from 315 ppmv (parts per million by volume) when we began systematic measurement in 1958 to about 380 ppmv today. There is also relatively little argument that preindustrial levels were about 280 ppmv. There is no disagreement that CO_2 is a gas with important absorption bands in the infrared.

There is agreement that at the level of fractions of a degree, the earth's global mean temperature is always varying, and there is widespread agreement (though with appreciably greater uncertainty) that over the past century there has been net warming of between 0.5 and 0.75° C (depending on which analysis one uses). This warming has, as far as anyone can tell, been irregular, with warming between 1920 and 1940, modest cooling between about 1940 and the mid-70's, warming between about 1976 and the early nineties, and little of either since.

Even the U.N.'s Intergovernmental Panel on Climate Change (IPCC)

---

* Professor Lindzen's presentation at the conference focused on the science of global warming, particularly "points of agreement vs. points of alarm." As with some of the other presentations, we were unable to convert his many slides into a format suitable for print. However, we are pleased to include this essay, in which he offers his perspective on the scientific and public debate over global warming. It was previously published by the Cornwall Alliance and in AIER's *Research Reports*.

acknowledges that greenhouse forcing is currently about three quarters of what one would expect from a doubling of CO_2, and yet we have seen much less warming at the surface than models project—even with models that have oceans which are supposed to delay the response.

Here the argument amounts to one between those like me, who think that the most likely reason for the discrepancy is that models are exaggerating the response, and those who think the models are correct, but that aerosols have cancelled much of the warming. However, even the IPCC acknowledges that our confidence in the aerosol cooling is low.

Agreement goes even further: there is general agreement that the famous 'blanket' picture of the greenhouse effect that Gore likes to present is, in fact, misleadingly wrong. Rather, the real greenhouse climate effect requires most warming to occur in the middle of the tropical troposphere (cooling at the surface is mainly by motion systems, with the heat deposited in the middle of the troposphere where it is then radiated to space), and as a recent report of the National Research Council notes, warming trends at this level in the tropics appear to actually be even smaller than at the surface.

For me personally, I find that the low climate sensitivity is consistent with my research on cloud feedbacks and other matters, but when it comes to current research one doesn't normally seek general agreement.

So where is there significant disagreement?

The main focus of disagreement has remained much the same since I first went public with my objections to catastrophic claims in 1988. (It is sobering to realize how long we have been told by environmental groups like the Union of Concerned Scientists that the end of the world as we know it is imminent due to global warming.) At that time, I felt confident, on the basis of my own research over the previous decade or more, that our knowledge didn't warrant these claims.

Given the long-term nature of climate, it should not be surprising that there is little reason to change this position. Nevertheless, it has, since the 1980s, led to an important disagreement with some of my colleagues over whether our present limited knowledge warrants deep concern or not. I, personally, don't think so, but I respect my colleagues' right to feel otherwise. This difference is distinct from the issue of whether concern is tantamount to feeling that specific actions are warranted. Most of my colleagues would agree, for example, that Kyoto is merely symbolic with little potential for affecting climate. Some favor other approaches, but I think there is widespread acknowledgment that with presently known or anticipated technology there is little that one can do to significantly cut greenhouse gas levels, and even less that one can do to significantly reduce radiative forcing by greenhouse gases (which, in the case of CO_2, goes up much more slowly than the level of CO_2 itself).

There are, of course, some who feel that warming concerns are a good excuse for implementing their pet energy policies. Here, I share with the late Roger Revelle (whom Gore points to as his mentor in this area) the view that current evidence does not warrant any drastic actions that cannot be justified independently of climate concerns.

Given my views, I am happy to be at an institution like MIT. At least most people at MIT are sufficiently technically savvy to appreciate the arguments involved with this issue. In the world at large, the situation is certainly different. No scientific issue has likely ever been as politicized as this one. Global warming has for about 20 years been a major focus of environmental advocacy groups and their political allies. In the last two years, they have greatly expanded their efforts to spread alarm to the public at large, including elementary school children, who lack any ability to understand the issue and are apparently suffering an appreciable degree of anxiety.

In any marketing effort, it is useful to offer the objects of the propaganda something that they value. In the present instance, they are offered at least two such benefits. First, they are given a sense of virtue: simply by changing light bulbs or (for the wealthier) buying a Prius or even by paying some outfit an indulgence to cancel their carbon footprint, they are made to feel that they are saving the world. Second, their intellectual insecurity when confronting such a complex issue is relieved by being told that all scientists agree with whatever propaganda they are fed. Under the circumstances, they are made to feel that in going along with the propaganda, they are displaying intelligence, and acquiring the right to consider anyone who does not as being either stupid or hopelessly corrupt.

Thus, the existence of questions about the validity of the global warming alarmism threatens both their virtue and their intelligence, and it should not be surprising that the response to such threats can be emotionally intense.

However, judging from my email, a great many people are beginning to resent being exploited in this manner. I fully expect that this latter group will eventually be vindicated, and that alarm over global warming will go the way of the Club of Rome forecasts for hunger and Y2K (not to mention the fears over global cooling of just 30 years ago).

In the meantime, there is growing pressure to take some action. Given that climate is likely to change as it always has regardless of man's activities, there is probably a case to be made to improve our capacity to adapt to changes of this sort. As a rule, the poor are more vulnerable to such changes than are the prosperous. Anything that will improve the economic state of the poor of the world will undoubtedly reduce their vulnerability. Anything that will inhibit this would have to be reckoned as immoral, and, unfortunately, some proposed actions seem to fall into this category. Obviously, any action that restricts the access of the poor to energy will fall into this category. But even innocent sounding policies such as mandating ethanol lead to increased food prices—not to mention adverse environmental consequences.

To buy publications or find out more about the
American Institute for Economic Research please contact us at:

American Institute for Economic Research
250 Division Street, PO Box 1000
Great Barrington, Massachusetts 01230

Phone: (413) 528-1216
Fax: (413) 528-0103

aierpubs@aier.org
www.aier.org

Call or visit our website for pricing.